BIOLOGICAL TECHNIQUES IN ELECTRON MICROSCOPY

BARNES & NOBLE INTERNATIONAL TEXTBOOK SERIES

BIOLOGICAL TECHNIQUES IN ELECTRON MICROSCOPY

CLINTON J. DAWES
University of South Florida

BARNES & NOBLE, INC., NEW YORK
Publishers • Booksellers • Founded 1873

© Copyright, 1971
By Barnes & Noble, Inc.

All rights reserved. No part of this book may be reproduced or utilized in any form or by any means, electronic or mechanical, including photocopying or recording, or by any information storage and retrieval system, without permission in writing from the Publisher.

Philippines Copyright, 1971
by Barnes & Noble, Inc.

L. C. Catalogue Card No. 70-153055
ISBN 389-00464-2

Distributed

In Canada by Methuen Publications, Toronto

In Australia and New Zealand by Hicks, Smith & Sons Pty. Ltd., Sydney and Wellington

In the United Kingdom, Europe, and South Africa by Chapman & Hall Ltd., London

Printed in the United States of America

To Dr. Flora Murray Scott,
Professor Emeritus, Department of Botany,
University of California (Los Angeles)

PREFACE

The use of the electron microscope as a tool in biological research has sharpened our concepts of cell structure and our understanding of the role of cell organelles in the overall structure of the cell. This research has both supported and altered previous information regarding cells obtained from studies utilizing the polarizing microscope and x-ray diffraction methods. The concepts of eukaryotic and prokaryotic cell organization have become common terms in modern biology. The techniques for preparing biological specimens for examination with the electron microscope are very new and are continually changing. Thus, there exists a demand for a description and discussion of standard preparative techniques used in electron microscopy. This book fulfills that need.

This text grew out of a course that I gave for several years in ultrastructural techniques. The original lectures have been expanded to include the following major techniques for the preparation of biological material: *fixation* including a discussion of the major fixatives and buffers used, *dehydration, embedding* including a number of common plastics and embedding procedures, *sectioning, knife manufacture, staining* for the electron microscope, *replication,* and *particulate specimen preparation.* Finally, a chapter on photographic techniques has also been included. Five appendices follow the text; their titles are "Commercial Sources and Chemical Checklist," "Chemistry of Epoxy Resins," "Sample Preparation Schedule," "Localization Techniques in Electron Microscopy," and a "General Reference List" with comments.

This text should serve as both a laboratory manual for students working under a professor and those beginning to use the electron microscope on their own, and also a supplementary reader for biologists who wish to become acquainted with research applications of the electron microscope. There appears to be a real need for a concise, standard procedures for the preparation of biological specimens. This book is therefore intended for the novice in electron microscopy who needs some infor-

PREFACE

mation not only on procedures, but also needs a recommended program for preparation techniques of biological material. Thus, each procedure is first discussed and then specific formulae or methods are given. In Appendix III, a complete step-by-step program is offered which may be followed by the investigator.

Since I feel that botanical material has been neglected in the previous texts, the subtle differences in the preparation of plant, animal, and bacterial materials are clearly pointed out. In the past few years a number of texts dealing with techniques for preparation of biological material for electron microscopy have been published (Kay, 1961; Pease, 1964; Mercer and Birbeck, 1966; Sorval, 1967; Sjöstrand, 1967; Juniper et al., 1970; and Wischnitzer, 1970). I have used a number of these books in my courses but have found them to be inadequate. This is because in some cases new techniques have become available since the publication of the texts. In addition, most of these texts have an orientation toward preparation of higher animal tissue. Finally, a few of these texts are simply small recipe books, without discussive or descriptive material.

Since this book is primarily a techniques book, numerous procedures are included which were obtained from the scientific literature, personal communication, and research experience. To acknowledge all who have aided in obtaining or refining the techniques would be impossible. Special thanks must go to Dr. Allan Wardrop, Monash University, Victoria, Australia, and to Dr. Flora Murray Scott, University of California at Los Angeles, where the author learned basic techniques in electron microscopy. The numerous students and professors who have aided me in my research and searches through the literature cannot be singled out by name, but their influence is obvious throughout the manuscript. I wish to thank Dr. Myron Ledbetter, Brookhaven National Laboratory, for constructive comments and criticism of the manuscript; and Dr. Robert Long for financial aid in preparation of the manuscript. Special appreciation for help in preparation of the manuscript and plates must go to Mrs. Lillian Stark, Mr. Harry Calvert, and Mrs. Joanne Davis, University of South Florida, Tampa. The photographs of various demonstrations were done with the professional aid of Mr. Harry Greer, a graduate student, and the Photography Department at the University of South Florida. Finally, the author assumes full responsibility for the contents of this text as to accuracy, although without such essential aid this manuscript would not have been possible.

Clinton J. Dawes

CONTENTS

Preface   vii

1   ELECTRON MICROSCOPY   1
   1.1 Introduction   1
   1.2 An Historical Synopsis of Electron Microscopy   3
   1.3 Electron Microscope Construction   4
   1.4 Resolution—A Brief Explanation   8
   1.5 Preparation of Specimens for Electron Microscopy   9
   1.6 Design of an Electron Microscopy Laboratory   13

2   FIXATION   17
   2.1 Introduction   17
   2.2 Chemical Fixation   17
   2.3 Chemical Fixatives   22
       Buffer   22
       Tonicity   22
       Temperature   23
       Duration of the fixation process   23
       Rinsing   23
       Presence of divalent ions   23
       Types of chemical fixatives   23
       Osmium Tetroxide (OsO_4)   26
         Fixation procedures   28
         Buffer formulae for osmium tetroxide   29
         (1) Palade (1952), Veronal-acetate buffered osmium (1%)   29
         (2) Dalton (1955), Chrome-osmium fixative (1%)   31
         (3) Millonig (1961b), Phosphate buffered osmium (1%)   31
         (4) McClean and Cook (1952), Sorensen's phosphate buffer and osmium (1%)   33
         (5) Kellenberger *et al.* (1958), Veronal-acetate buffered osmium (1%)   33
       Aldehydes (HCHO)   34
         Fixation procedures   35
         Buffer formulae for aldehydes   35
         (1) Pease (1964), Phosphate buffered formaldehyde (10%)   36

CONTENTS

        (2) Sabatini *et al.* (1962), Cacodylate buffered glutaraldehyde (5%)   36
        (3) Holt and Hicks (1961b), s-Collidine buffered glutaraldehyde (5%)   38
        (4) Millonig (1961b), Phosphate buffered glutaraldehyde (10%)   39
        (5) Gibbs (1962), Veronal-acetate-salt buffered glutaraldehyde (5%)   39
        (6) Ramus (1969), Cacodylate-sucrose buffered glutaraldehyde (5%)   40
        (7) Luft (1959), Cacodylate buffered acrylic aldehyde (5%)   41
        (8) Karnovsky (1965), Cacodylate buffered combination: glutaraldehyde-formaldehyde (3% each)   42
        (9) Dawes (1969), Phosphate buffered glutaraldehyde-acrylic aldehyde (3% each)   42
    Permanganates (MnO_4)   43
        Fixation procedures   43
        Buffer formulae for permanganate   44
        (1) Luft (1956), Veronal-acetate buffered permanganate (5%)   45
        (2) Mollenhauer (1959), Unbuffered permanganate (5%)   45
        (3) Dawes and Rhamstine (1967), Veronal-acetate-salt buffered permanganate (0.6%)   45
        (4) Dawes and Rhamstine (1967), Sea water buffered permanganate (0.6%)   47
  2.4 Mechanical Fixation   46
    Freeze Drying   46
    Freeze Etching   47

3   DEHYDRATION   49
  3.1 Introduction   49
  3.2 Dehydration Schedules   51
    Ethanol-Propylene Oxide—Slow Schedule   52
    Ethanol-Propylene Oxide—Rapid Schedule   52
    Acetone Dehydration   52
    Infinite Dilution   53
    Other Series   53

CONTENTS

4	**EMBEDDING**	55
	4.1 Introduction	55
	4.2 Handling of Plastics	56
	4.3 Infiltration	57
	4.4 Capsule Embedding	57
	4.5 Flat Embedding	61
	4.6 Particulate Specimen Embedding (Pre-embedding)	61
	Agar Pre-embedding 64	
	Bovine Albumen Pre-embedding 64	
5	**PLASTICS**	65
	5.1 Introduction	65
	5.2 Methacrylate	65
	Plastic Mixture and Embedding 66	
	5.3 Epoxy Resins	67
	Epoxy Resin Formulae 68	
	(1) Glauert and Glauert (1958), English araldite 68	
	(2) Luft (1961), American araldite 68	
	(3) Cargille (1962), American araldite 69	
	(4) Luft (1961), Epon 70	
	(5) Spurr (1969), Epon 71	
	(6) Mollenhauer (1963), Epon-araldite mixture 72	
	(7) Freeman and Spurlock (1962), Maraglas mixture 73	
	(8) Ryter and Kellenberger (1958), Vestopal 74	
	5.4 Removal of Epoxy Resins	75
6	**BLOCK TRIMMING AND KNIFE MAKING**	77
	6.1 Block Trimming	77
	6.2 Knives	79
	Metal Knives 79	
	Diamond Knives 80	
	Glass Knives 84	
7	**SECTIONING AND THE ULTRAMICROTOME**	93
	7.1 Theoretical Considerations	93
	7.2 Historical Synopsis of the Ultramicrotome	94
	7.3 Current Ultramicrotome Models	96
	7.4 Sectioning	97
8	**PREPARATION OF SPECIMEN GRIDS**	107
	8.1 Types of Grids	107
	8.2 Grid Cleaning	108
	8.3 Supporting Films—Plastic	108

CONTENTS

        Method I   109
        Method II   112
        Method III   113
        Preparing films with holes   113
  8.4 Supporting Films—Carbon   113

9 HIGH VACUUM EVAPORATION, REPLICATION, AND PARTICULATE SPECIMENS   115

  9.1 The Evaporator   115
        Carbon Evaporation   115
        Metal Shadowing   115
  9.2 Replication   118
        Henderson (1969), Cellulose-acetate replica technique   122
        Wardrop (1964), Wet replica technique   123
  9.3 Particulate specimens   126
        Fragmentation   126
        Mounting   127

10 STAINING   131

  10.1 Introduction   131
  10.2 Penetration of Stains   131
  10.3 Staining Procedure   132
  10.4 Stain Formulae and Reactions   133
        Lead Stains   133
          (1) Reynolds (1963), Lead hydroxide chelated with citrate   140
          (2) Millonig (1961), Lead hydroxide, chelated with tartrate   141
          (3) Dalton and Zeigel (1958), Lead hydroxide chelated with subacetate   141
          (4) Dalton and Zeigel (1958), Lead hydroxide chelated with acetate   142
          (5) Karnovsky (1961), Lead hydroxide   142
          (6) Watson (1958), Lead hydroxide   143
          (7) Converse (1969), Lead hydroxide   143
        Uranyl Stains   144
          (1) Watson (1958), Uranyl acetate   144

CONTENTS

       (2) Uranyl nitrate   145
       (3) Uranyl acetate-lead citrate double stain   145
   Phosphotungstic Acid Stains   145
   Permanganate Stains   145
   Vanadyl Salt Stains   146
       (1) Vanadium sulfate   146
       (2) Vanadotomolybdate   146
   Ruthenium Red   146
10.5 Negative Staining   146
   Bovine serum base, phosphotungstic acid stain   147
   Starch base, phosphotungstic acid stain   147
   Sucrose base, phosphotungstic acid stain   148
   Uranyl acetate stain   148
10.6 Thick Section Staining   148
   Toluidine blue   148
   Methylene blue-azure II   149
   Basic fuchsin   149
   Crystal violet   150

11 PHOTOGRAPHY   151

11.1 Introduction   151
11.2 Chemistry of Photography   151
11.3 Exposure   152
11.4 Film Speed   152
11.5 Electron Recording (DQE)   153
11.6 Common Plates, Films, and Printing Papers   156
   Projector Slide Plates   156
   Kodak Electron Image Plates   157
   Dupont Cronar Sheet Film   157
   Kodak Electron Microscope Film   158
   Kodak Fine Grain Positive 35 mm Film   158
   Common Printing Papers and Processors   158
11.7 Darkroom Procedures   159
   Developing Negatives   159
   Printing and Enlarging   163

REFERENCES   167

APPENDIX I COMMERCIAL SOURCES AND CHEMICAL CHECKLIST   173

CONTENTS

I.1	Beem Capsules	173
I.2	Electron Microscope Companies	173
I.3	General Electron Microscope Supply Houses	173
I.4	Laboratory Chemicals—A Checklist	174
I.5	Plastics	175
I.6	Diamond Knives	176
I.7	Knife Makers	176
I.8	Ultramicrotomes	176
I.9	Plastic Film Materials	176
I.10	Specimen Grid Supply Houses	177
I.11	Special Sources of Osmium Tetroxide	177
I.12	High Vacuum Evaporator Companies	177

APPENDIX II CHEMISTRY OF EPOXY RESINS 178

II.1	Introduction	179
II.2	Epoxy Resin	179
II.3	Hardeners	180
II.4	Plasticizers	181
II.5	Accelerators	182
II.6	Curing Mechanisms of Anhydride-Epoxy Resins	183

APPENDIX III SAMPLE PREPARATION SCHEDULE 185

III.1	Introduction	185
III.2	Fixation	185
III.3	Dehydration	186
III.4	Embedding	186
III.5	Block Trimming and Grid Preparation	186
III.6	Sectioning and Staining	186

APPENDIX IV LOCALIZATION TECHNIQUES IN ELECTRON MICROSCOPY 189

IV.1	Introduction	189
IV.2	Immunochemical "Staining"	189
IV.3	Enzyme Localization	189
IV.4	Autoradiography	190

APPENDIX V GENERAL REFERENCE LIST 191

1 ELECTRON MICROSCOPY

Introduction 1.1

This book is a techniques book for the preparation of biological material for examination by the transmission type electron microscope. The goal of this book is to develop for students of biology, as well as investigators, an understanding of the use of the electron microscope. The instrument itself is relatively new and its possibilities in biology are immeasurable. The development and use of the electron microscope represents one of the most rapid advances of a research tool in the history of science. In comparison, the development of the light microscope took about 370 years, while the present-day electron microscope is the result of less than 50 years' development.

TABLE 1-1 Comparison of resolution capabilities of microscopes.

Unit of observation	Traditional units	SI units
Eye	ca. 0.1 mm	0.1 mm
Light (bright field microscopy)	ca. 0.2 μ	0.2 μm
Ultraviolet microscopy	ca. 0.1 μ	0.1 μm
Electron microscopy (normal transmission)	ca. 1 Å	0.1 nm

The importance of the electron microscope to biology can be seen by a comparison of resolution capabilities of several microscopes (Table 1-1). The light microscope extended man's visual observations to particles as small as 0.2 micrometers (μm). The electron microscope is capable of viewing particles (or rather molecules) the size of 0.1 to 0.2 nanometers (nm). The traditional units of measurements are given in Table 1-2 as well as the international units, or SI units, which will be adhered to in this text.

Since 1950, the electron microscope has been used for direct examination of cell structure down to the level of macromolecules. Thus, this instrument has bridged the gap between direct

1 ELECTRON MICROSCOPY

TABLE 1-2 Comparison of units of length for SI and traditional measurements.

SI (Système International d'Unités)			Traditional	
Unit name	Symbol	Fraction of meter	Unit name	Symbol
Meter	m	10^{-0}	Meter	m
Decimeter	dm	10^{-1}	Decimeter	dm
Centimeter	cm	10^{-2}	Centimeter	cm
Millimeter	mm	10^{-3}	Millimeter	mm
Micrometer	μm	10^{-6}	Micron	μ
Nanometer	nm	10^{-9}	Millimicron	mu
...	...	10^{-10}	Angstrom	Å

SI units: mm μm   nm
0.0000000000
Traditional units: mm μ mμ Å

examination by light microscopy (resolution *ca.* 0.2 μm, Table 1-1) and determination of molecular structure by x-ray diffraction and birefringent microscopy.

The discoveries resulting from electron microscopy have greatly influenced modern biology. A modern book in biology refers frequently to the concepts of virus structure, prokaryotic, and eukaryotic cell structure. With regard to virus structure, the electron microscope quickly yielded much new information. First, these organisms, being too small to be seen individually under a microscope (10 to 500 nm), were found to have a number of characteristic shapes and sizes. From this information, certain malignant types were then identified in tumor tissue. Secondly, the instrument provided an accurate method for counting virus particles. Finally, the electron microscope permitted investigators to determine the substructure of the protein coat and the process of viral infection of bacteria. A number of Nobel prizes have been awarded in this field, especially with regard to the transfer of genetic information.

Probably the greatest contribution to biology has been the use of the electron microscope in fine structural studies of cells of higher organisms. The chief difference between light and electron microscopic images of cells lies in the far greater amount of detail seen with the electron microscope. One of the important generalizations to emerge from the explosive advances in knowledge of cell fine structure concerns the ubiquity of complex membrane systems. The various organelles such as mitochondria, plastids, and Golgi bodies appear as sharply defined objects with intricate internal structure. A number of previously unknown organelles have also been found (lysosomes, lomasomes). This detail, of course, is seen because of the thousandfold increase in resolving power of the electron microscope, a topic to be covered later in this chapter. The fine detail has permitted biochemists, as well as biologists, to postulate how various physiological processes could be carried out in the respective organelles; that is, where the enzymatic systems might reside and how these complexes might be structured. The amount of basic information still to come from as new an instrument as the electron microscope is undoubtedly enormous.

An Historical Synopsis of Electron Microscopy 1.2

The first electron microscope with electromagnetic lenses was built by Knoll and Ruska in the years 1930 to 1933. Commercial electron microscopes were made by Siemens in Germany in 1939 and by the American firm RCA in 1941. The first electron micrographs or photographs of biological material were taken with the electron microscope as early as 1934 (a bacterial cell, whole mount). The main breakthrough, however, in the study of biological specimens did not come until 1952 and 1953. At this time, ultrathin sections were obtained using a modified histological microtome, good fixation with osmium tetroxide, and plastic embedding. The most successful techniques available up to that time had been whole mount preparations and replication. Today, electron microscopy finds many fields of application such as structural analysis in engineering, study of prism surfaces in physics, study and analysis of minerals in geology, detection and study of tumor-causing virus in medicine, and localization of enzyme systems in cells in biology. There are a number of major firms which manufacture transmission, standard voltage electron microscopes (JELCO, Akashi and Hitachi of Japan; Philips of Holland; AEI of England; Ziess and Siemens of Germany; RCA in the U.S.A., now being made by Forgflo). The resolution achieved by these instruments ranges from 1.5 to 0.1 nanometers (nm) with new types of electron microscopes being developed at a rapid pace. Some of the newer instruments include the scanning electron microscope (Echlin (1968), Thornton (1968)), the electron probe systems, and the high voltage electron microscopes (Cosslett (1967), Hama and Porter (1969), Ris (1969)). Major books and articles dealing with all aspects of electron microscopy and a number of excellent texts on the electron microscope can be found listed in Appendix V.

1 ELECTRON MICROSCOPY

1.3 Electron Microscope Construction

The electron microscope is a type of microscope and therefore conforms to the definition of a microscope which is "an optical instrument consisting of a lens or combination of lenses used for making enlarged or magnified images of minute objects."

Figure 1-1 shows a Philips EM 200 electron microscope with the major column components labeled, and Fig. 1-2 is a diagrammatic comparison of a light and an electron microscope. The major distinctions between the two optical instruments are the types of resolving media used (visible light for the light microscope and electrons for the electron microscope), the type of lens system utilized (glass lenses for the light microscope and electromagnetic lenses for the electron microscope), and the requirement for a vacuum in the electron microscope. A rather high vacuum is required since the free path of an electron is only 2.5 m in a vacuum of 10^{-4} torr (note that 1 torr is equal to 1 mm of mercury pressure and that 1 atmosphere of pressure equals 760 mm of mercury).

As shown in Fig. 1-2, the electron microscope is basically similar in construction to a light microscope except that it is inverted. The electron gun takes the place of the lamp of the light microscope as the source of illumination and the glass lenses are replaced by electromagnetic lenses (magnetic fields that focus the electron beam). Since electrons cannot be seen by the human eye, a viewing screen coated with a material which fluoresces is used (usually zinc sulfide crystals). The three basic lens systems —condenser, objective, and projector lenses—function as in a light microscope. In the electron microscope there is a mechanical stage on which the specimen is placed, as in the light microscope.

Figure 1-1 Photograph of a Philips EM 200 electron microscope. The labels on the various components are as follows: S is the electron source (gun); C the condenser lenses; O the objective lens; D the diffraction lens; I the intermediate lens; and P the projector lens. The specimen is inserted at the arrow to the right of the column and the viewing screen is directly below the projector lens. Two cameras are present in this instrument: a 35-mm camera (or optional 70-mm camera) just below the projector lens (35) and a $3\frac{1}{4}$- by 4-in. plate camera below the viewing screen (labeled 3 × 4).

CONSTRUCTION 1.3

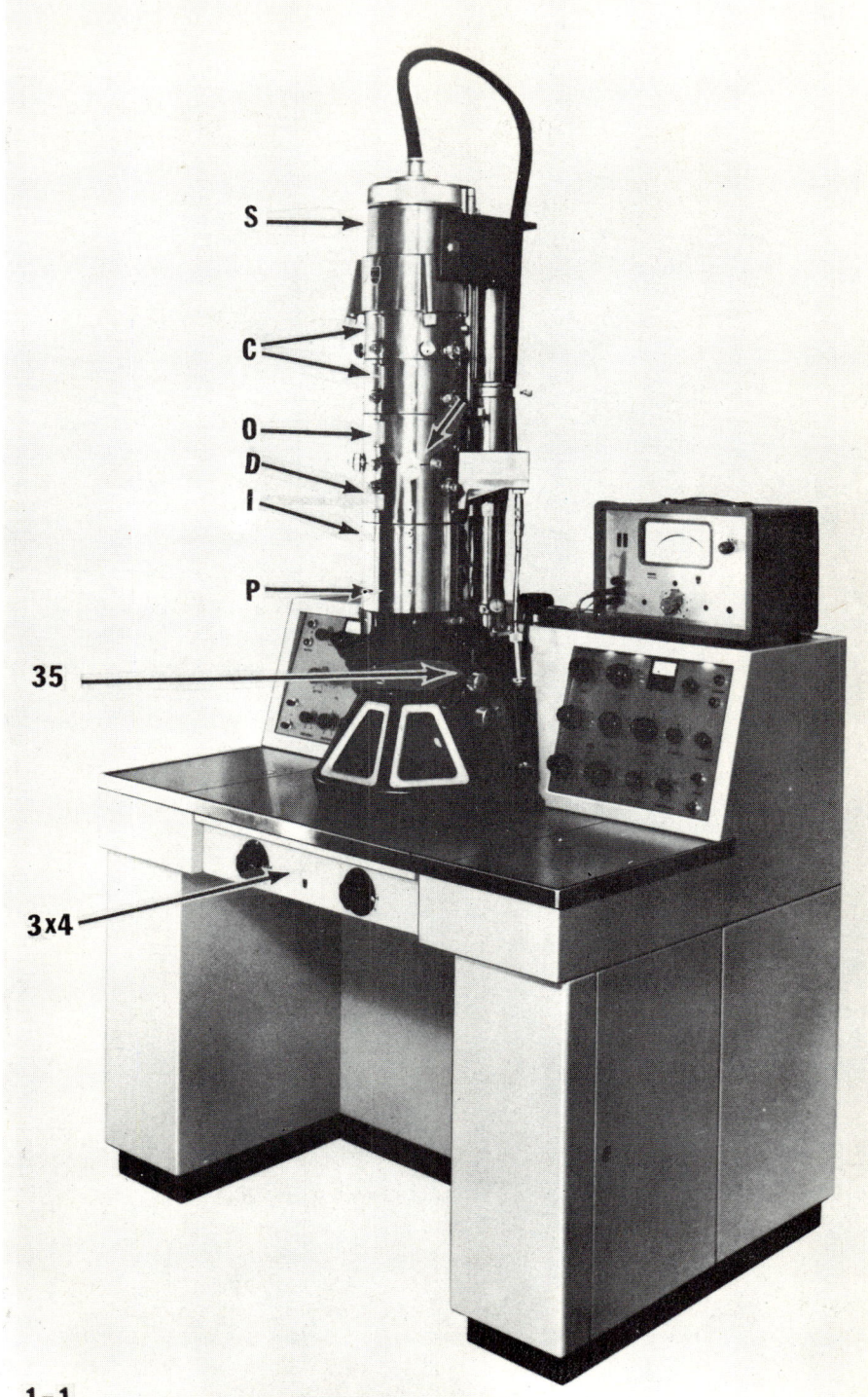

1 ELECTRON MICROSCOPY

Figure 1-2 A diagrammatic comparison of a light and an electron microscope. The electron microscope is inverted to permit similar positioning of the various lens systems. The two major differences illustrated in this diagram are the type of lenses (electromagnetic vs. glass) and the type of medium used (electrons vs. light).

Figure 1-3 Diffraction waves produced by a source of light and passed through a pinhole aperture. [Modified from Wischnitzer (1962).]

CONSTRUCTION 1.3

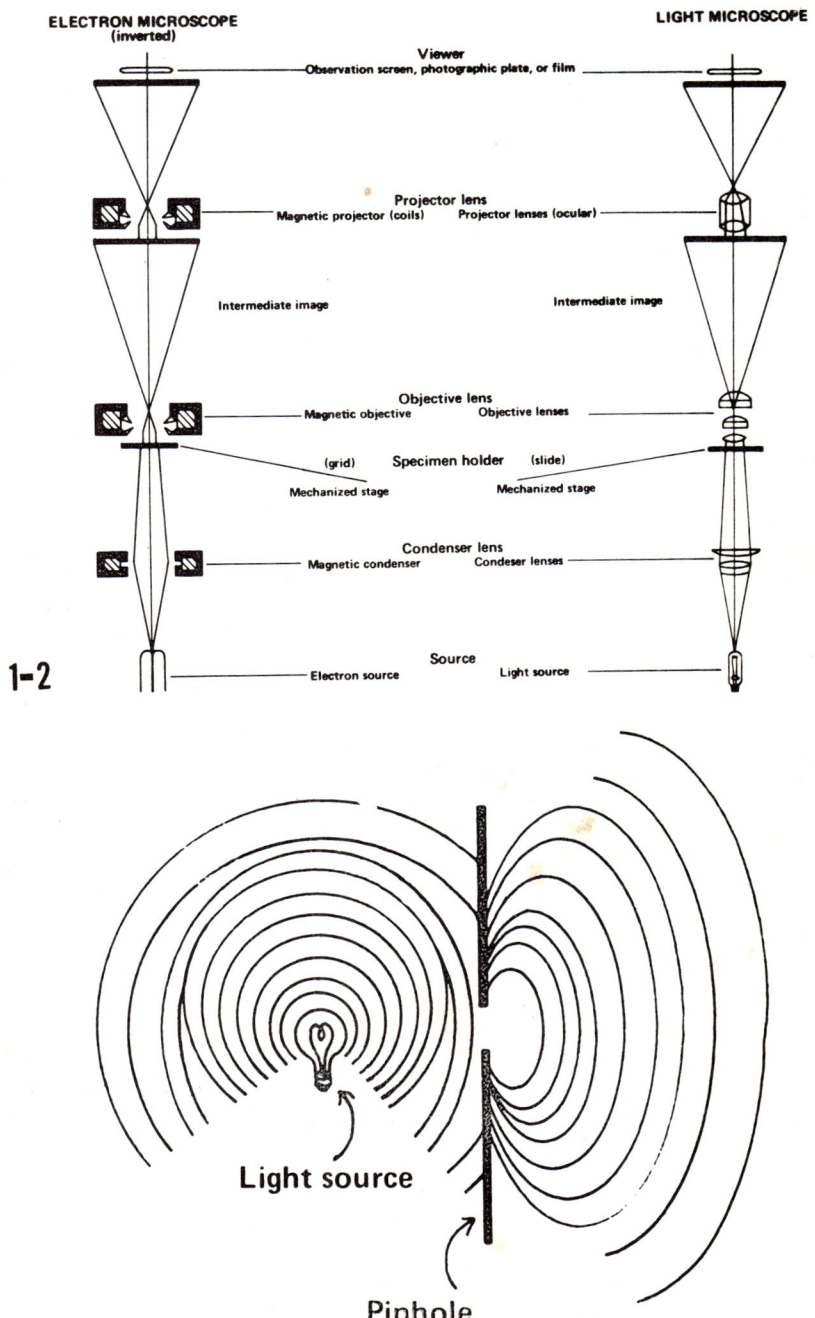

1 ELECTRON MICROSCOPY

1.4 Resolution—A Brief Explanation

Probably the most important characteristic of any optical instrument is its *resolving power*. We can define resolution as the smallest distance two points can be separated and yet be observed as distinct entities. Table 1-1 gives the resolving power of light, ultraviolet, and electron microscopes. The light microscope, according to this chart can just resolve objects that are 0.2 μm apart, while the electron microscope can resolve units that are only 0.1 nm apart. In other words, the electron microscope has almost a thousandfold increase in resolving power.

A number of features of the electron microscope contribute to the attainment of such a high resolving power and probably the most important of these is the extremely short wavelength (λ) of high-speed electrons. To explain this, the phenomenon of *diffraction* must first be understood. Diffraction is the bending or spreading of light into an area behind some blocking object through which the light waves have passed. The blocking object might be a sheet of paper with a pinhole or aperture which permits a portion of the light beam to pass through it (Fig. 1-3). The spot of light resulting from this aperture when directed through a lens and onto a screen gives an image that is not sharp, but instead shows a series of diffraction or "Airy" rings (Fig. 1-4). Likewise, the same effect occurs around the edge of the limiting apertures and lenses, as well as all objects focused by a lens system in microscopy. In other words, these diffraction patterns are *inherent* in optical microscopy whether the resolving medium is light or electrons.

Diffraction is mentioned because of its influence on resolution. Diffraction is in fact the limiting feature for resolution in light microscopy. Based on work by others (Sir George Airy and Lord Rayleigh) Ernst Abbe, working for Ziess in Germany during the nineteenth century, formulated a relationship between the wavelength (λ), the aperture opening of the microscope (α) (Fig. 1-5), and the medium through which the wave passes (n). This equation is now known as Abbe's equation and is as follows:

$$d = \frac{0.612\lambda}{n \sin \alpha}$$

where 0.612 is a constant (varying from 1.22 to 0.61 depending on the size of the illuminating aperture); λ the wavelength of the radiation used to form an image; n the index of refraction of the medium through which the wave passes (free space is 0, while

PREPARATION OF SPECIMENS 1.5

water and oil are higher, respectively, 1.2 and 1.6); α the aperture angle of the illuminating wave (the sine of this angle can never equal more than 1, $\sin 90° = 1$); and d is the diameter of the first dark diffraction ring.

The important feature of this equation is that it describes the magnitude of diffraction and can be used with relation to points in the specimen plane. In other words, two points in the specimen (e.g., unit membranes of mitochondria) will each produce a diffraction disc (Airy disc) in the image plane (e.g., image screen or negative). The size of the diffraction rings from the points in the specimen plane produced by the optical instrument on the image plane will be dependent on wavelength, index of refraction, and aperture angle. If the diffraction discs produced by the two points do not overlap, then they will be resolvable as two separate points in the object (Fig. 1-6). If, however, the diffraction discs do overlap, then the image will appear as if there were but one point (Fig. 1-6). That is, the instrument cannot "resolve" the two points as separate units. Through d in Abbe's equation, we can determine how small a distance two points must be apart and yet still be resolvable as distinct entities. This distance d, in turn, is determined by the factors in the equation. If we look at wavelengths of different types of radiation, we can see why Abbe's equation of diffraction is so important. An electron accelerated by 50 kV will have a wavelength of only 0.005 nm. For comparison, we note that the wavelength of visible light varies from about 400 to 800 nm (blue to red light). Therefore by Abbe's equation, we deduce that the shorter wavelengths of electrons and their greater speed will produce smaller diffraction rings and permit greater resolution. The higher the acceleration of an electron, the shorter the resulting wavelength of the electron. This is why electrons subjected to higher accelerating voltages are now being used (5 million volt instruments are now in use). This very interesting subject can be pursued further by referring to textbooks on the theory and optics of electron microscopy (see General Reference List in Appendix IV).

Preparation of Specimens for Electron Microscopy 1.5

The preparation of specimens for electron microscopy is greatly influenced by the low penetrating power of electrons, the vacuum requirement, and the use of high accelerating voltages that reduces contrast of the image.

1   ELECTRON MICROSCOPY

Figure 1-4   A diagram of Airy or diffraction rings produced on a screen from an arrangement shown in Fig. 1-3 and with a lens interposed. The labels are: s the source; p the pinhole aperture; a the circular limiting aperture; l the biconcave lens; sc the viewing screen; and d the Airy disc. [Modified from Wischnitzer (1962).]

Figure 1-5   The path of light in a microscope showing the half aperture angle. The labels are a_c the aperture of the illumination which reaches the specimen plane; s the specimen plane; a_o the aperture angle of light traveling from specimen plane to objective lens. [Modified from Wischnitzer (1962).]

PREPARATION OF SPECIMENS 1.5

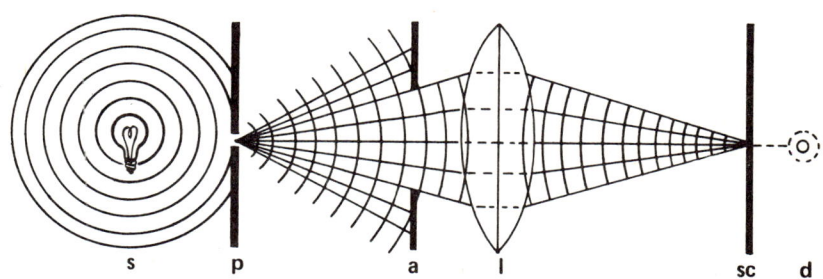

1-4

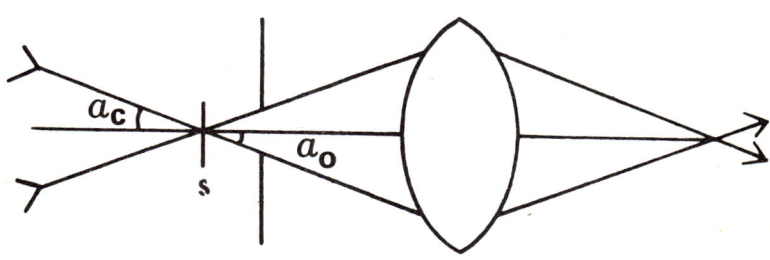

1-5

1 ELECTRON MICROSCOPY

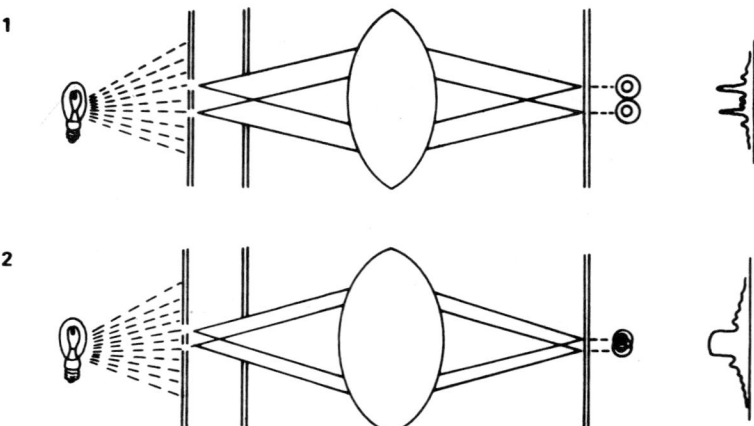

Figure 1-6 The two diagrams demonstrate the resolution of two apertures based on the sizes of the respective diffraction discs. Diagram 1 shows that if the apertures are far enough apart they will be resolved as two points (see intensity peaks at far right). Diagram 2 shows that if the apertures are too close together the diffraction discs will overlap and therefore not permit resolution (intensity peaks are combined). [Modified from Wischnitzer (1962).]

Specimens must be very thin, ranging from 20 to 100 nm. This is necessary because of the poor penetration of electrons accelerated at 40 to 100 kV. A second reason, partially dependent on the great depth of field of the transmission electron microscope, is that thicker sections tend to produce blurred images on the photographic plate. The electron microscope, operating at 50 kV has a depth of field of around 1 μm. This means that in a $\frac{1}{2}$ μm thick section all the structure contained is in equal focus on the photographic plate. Such a condition results in superimposition of membranes, causing loss of resolution.

The requirement of a stable, high vacuum for electron acceleration necessitates that the biological material be properly fixed, dehydrated, and then embedded in a suitable medium such as epoxy plastic.

Finally, because an intense, high-speed electron beam must be focused on the specimen, the embedding medium must be thermostable. The combination of a dense embedding medium and high-speed electrons reduces the contrast in the image. This problem is further enhanced by the fact that biological material consists primarily of carbon, hydrogen, oxygen, and nitrogen atoms that will not deflect electrons. Consequently, the addition of heavy atoms is required at fixation and after sectioning to increase electron deflection and, therefore, improve contrast.

Biological material for electron microscopic studies usually takes one of the three following forms: whole mount preparation, replication of surfaces of specimens (including freeze etching), or ultrathin sectioning of properly prepared material. Ultrathin sectioning requires the greatest effort in specimen preparation and is presently the most useful method for study of cell structure in electron microscopy. This procedure will be featured in this book and it requires a number of separate techniques including: fixation (chapter 2), dehydration (chapter 3), embedding (chapters 4 and 5), sectioning (chapter 7), staining (chapter 10), and specimen grid preparation (chapter 8). These techniques are presented in the same order that the investigator proceeds to produce ultrathin sections. In addition, there is an Appendix (III) presenting a recommended procedure for preparing specimens for sectioning. Whole mount preparation and replication are most effective, especially in the study of particles, bacteria, and viruses, and in the study of cell surfaces. These details are covered in chapters 9 and 10. Because photographic techniques are used consistently in electron microscopy, chapter 11 is included.

Design of an Electron Microscope Laboratory 1.6

The components of a biologically-oriented ultrastructure laboratory depend on the amount of funds available. The purpose here is to outline the minimum components needed to provide for a variety of biological research projects. Such a laboratory can be as small as a basic three-room suite—a large preparation room, a smaller room containing the electron microscope, and a small darkroom (Sell and McMaster, 1963).* If possible, it is advisable to have a separate room for ultrathin sectioning and two darkrooms, one for developing electron microscope negatives and the other darkroom for printing.

---

* Specific references are found in the Reference section, which follows chapter 11.

1 ELECTRON MICROSCOPY

The preparation room contains the necessary chemicals, glassware, and equipment that will support biological research. A large fume hood and a refrigerator with a fairly large freezer are essential. Normally, two embedding ovens operating at around 80 and 48 degrees centigrade (°C), respectively, are included. Other standard items include a light microscope with camera and calibrated ocular micrometer (phase-contrast objectives, if possible), a dissecting microscope with trimming block holder and calibrated ocular micrometer, a pH meter, balances, centrifuge, glass-knife making tools, a small vise, and normal working tools such as hammer, files, screwdrivers, and allen wrench and ratchet sets. In addition, specialized equipment such as an ultramicrotome and a high vacuum evaporator can be placed in this room.

In order to reduce contamination, it is recommended that regions of the preparation room be specifically designated for activities such as staining, knife making, block trimming, dehydration, embedding, and sectioning. Specialized equipment should be carefully placed. For example, since the ultramicrotome is sensitive to vibrations, it should be mounted on a sturdy table top or on a modified table (see chapter 7). Furthermore, the cutting portion of the microtome must be shielded from air currents that cause expansion or contraction of its advance mechanism. If possible, sectioning should be done in a separate room.

The specifications of the electron microscope room are usually given by the manufacturer and include the minimal size, air conditioning requirements, electrical outlets, and water supplies. The control of the room lighting can be arranged so that it is regulated from the microscope console, thus saving the operator a trip to the wall switch.

The electron microscope is very often used to take pictures or micrographs of the material placed in view. Thus, the darkroom should be adjacent to the electron microscope room to facilitate the loading and transfer of cameras from the microscope to the darkroom. Since more than one person may require the use of such facilities, it is important, if at all possible, to keep the two functions of developing and printing in separate darkrooms. If only one darkroom is available, it should be equipped with a sink (with varying water temperature to help control the development process), light-tight storage cabinets, "safe" lights with appropriate filters, timer, enlarger, dryer, normal development-printing trays or tanks, and a sufficient work area.

LABORATORY DESIGN 1.6

Finally, a general word about the use of the laboratory. Ideally, the use of the electron microscope should be restricted to one or two individuals. The instrument is rather sensitive and much effort is needed to maintain it at its maximum resolution. It is recommended, if a large number of investigators are using the same instrument, that a technician be hired. His job is to operate the electron microscope, to act as a watchdog, and to institute and enforce strict rules as to the operation and control of the microscope, darkroom, and preparation room. The cost of such a technician can readily be recouped through a more efficiently run electron microscope laboratory by avoiding damaging and costly errors.

2 FIXATION

Introduction 2.1

The purpose of fixation is to alter the tissue constituents in a way which renders them no longer susceptible to autolysis or decay through microbial action and to make them immune to distortion from subsequent treatment. Fixation should make as permanent and as true a picture of cell structure as is possible. Fixation should represent a "frozen slice of life," the ideal goal of the cytologist.

Two major types of preservation of biological material for electron microscopy are known: chemical fixation and mechanical fixation. Prior to the actual fixation, a most important step is tissue preparation and handling which will be presented below.

Chemical Fixation 2.2

Every section of biological material viewed in the electron microscope represents an investigator's time. Consequently, it is imperative that every step taken toward the viewing of sectioned material be executed with care. The first step is the preparation of the tissue for fixation. The following section will present specific fixation formulae.

Some general recommendations for the preparation of the specimen, whether it be eukaryotic, prokaryotic, plant, or animal, are:

1. Have all necessary tools ready for use—scalpel, razor blades, toothpicks, vials, fixative, and ice baths (see Fig. 2-1).
2. Use only sharp cutting instruments; dull razor blades or scalpels will only damage the tissue.
3. The size of the specimen for fixation varies according to the fixative used. Penetration of different fixatives varies greatly. However, one rule is that at least one dimension be a millimeter (mm) or less in thickness.
4. Temperature is critical (for example, with permanganate). Therefore, the handling of the tissue should be conducted under controlled temperatures.

2 FIXATION

Figure 2-1 A demonstration of materials needed for fixation. The fixatives are stored in glass stoppered bottles and properly labeled. A sufficient supply of complete buffer is also present to permit thorough washing of the fixed tissue. Rough trimming can be done in a petri dish cover placed in the fixative. Final trimming or mincing of the tissue should be done in a drop of fixative on a cardboard sheet. Toothpicks can be used for transferring or holding the tissue. Tweezers, dissecting needles, a sharp razor blade, or scalpel should be available. An ice bath may be required for some fixatives.

Figure 2-2 A close up view of the mincing process with a razor blade. The tissue (below the hand) is held in place with a toothpick and the razor blade is used in a rapid chopping motion.

Figures 2-3 through 2-8 Electron micrographs of ultrathin sections of the pituitary gland of *Poecilia latipinna* (sailfin molley) a common fish of Tampa Bay prepared by Mr. Harry Grier, University of South Florida. Each figure represents a different fixation, and all were stained in an identical manner (uranyl acetate, 30 min followed by lead citrate for 15 min). The unit mark on each micrograph represents 1 mm.

Figure 2-3 Fixation with 1% osmium tetroxide in Millonig's phosphate buffer. The mitochondria (M) are well fixed although the nuclear membrane (N) appears to be swollen. The secretion droplets within the cells are also well stained with osmium. The vesiculization of the membranes is believed to be caused by the rather low tonicity of the buffer and not the fixative ($\times$ 6400).

CHEMICAL FIXATION 2.2

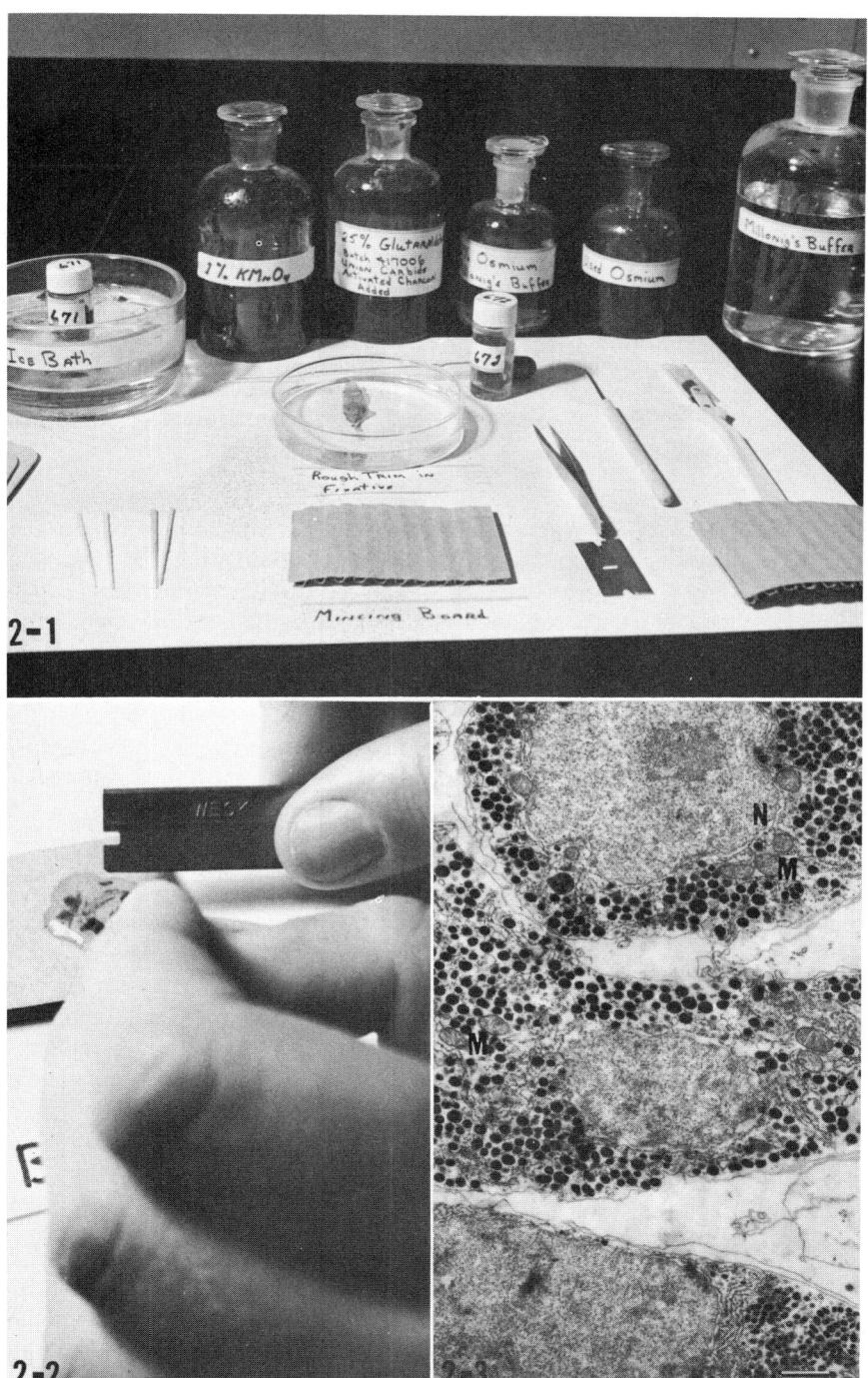

2 FIXATION

5. Use toothpicks to hold specimens down while dicing and to transfer tissue segments to vials (Fig. 2-2). This prevents mechanical damage to the tissue which might arise if forceps are used.
6. Techniques for handling and preparation of unicellular material for fixation can be found together with descriptions of the fixative solutions in chapter 4 under pre-embedding (particulate specimen embedding).

For larger animals, it may be deemed best to fix tissue by *in situ* fixation or by perfusion fixation. There are a number of techniques outlined by Pease (1964) for topical and *in situ* fixation and by Sjöstrand (1967) for fixation by perfusion. Here, a simplified procedure will be described for each technique. With regard to the techniques, the investigator should pay strict attention to the control of pH, tonicity, and temperature of the fixative solutions used for different tissues and animals.

Prior to *in situ* fixation of an organ, the animal may be anesthetized and placed in a dissecting tray. After a neat, straight incision is made to bare the organ, the animal is then arranged to best show the organ to be fixed. The capsulating membranes around the organ are pulled away with fine-pointed tweezers and the organ is rapidly flushed with fixative. Oozing blood is removed by reflushing the organ with a dispensing pipette. The fixative will soon stop any further bleeding. As recommended by Pease (1964) the excess fixative and body fluids should be drained off the organ onto surrounding cotton pads, which should be changed frequently.

During fixation, the organ should be kept moist with fresh fixative. A thin layer of cotton is gently laid over the exposed organ and then covered with a sheet of parafilm. A Pasteur pipette, filled with fresh fixing solution, can be stuck through the parafilm to touch the cotton over the organ. This pipette can replenish the fixative solution in the cotton covering. After completion of *in situ* fixation, the organ is removed from the animal by dissection and placed in a petri dish with fresh fixative. The organ is cut up and fixation is continued as described below.

Fixation by perfusion uses the vascular system of the animal to distribute the fixative throughout the organ. The investigator first locates the proper artery by dissection of the anesthetized animal. In some cases, he may only need to inject the fixative solution into the artery via a syringe to obtain good fixation. In other

CHEMICAL FIXATION 2.2

cases, the perfusion apparatus consists of a bottle with perfusate (fixative plus buffer) and rubber tubing connecting the bottle of fixative to a pipette which is inserted into the artery. Fixation by perfusion requires accurate adjustment of the fixing solution with respect to temperature, pH, tonicity, and ion composition of the solution for otherwise vasoconstriction will prevent the fixative solution from reaching the tissue and fixation artifacts will result. The fixative solution may vary from one organ to another. Sjöstrand (1967) presents a variety of fixing solutions for perfusion of brain, kidney, and liver. Finally, if artificial respiration is to be used, Sjöstrand recommends at least 5% CO_2 in the gas to prevent possible vasoconstriction of the artery.

Regardless of whether material is fixed *in situ,* by perfusion, or simply cut off and placed in the fixative, the following procedure should be employed. When organs are fixed *in situ* or by perfusion, this procedure is still recommended to insure adequate fixation in the central region of the organ and good dehydration and embedding.

1. Place the excised material in a small petri dish with fresh fixing solution.
2. With a sharp razor blade or scalpel, cut off a small section and transfer it with a flattened end of a toothpick to a cutting board (small sheet of waxed paper, dental wax plate, or cardboard sheet) which has a fresh drop of fixing solution on it (Fig. 2-2).
3. Slice off sections about $\frac{3}{4}$ mm thick and transfer these sections with the flat end of a toothpick or shaved applicator stick into labeled vials containing fresh fixative.
4. If temperature is a factor in fixation, the entire procedure can be done in a small dish placed in an ice bath (Fig. 2-1).

Some fixatives such as osmium tetroxide and potassium permanganate react with the razor blade and, consequently, the blades should be changed often. Because of the danger of fumes of aldehydes and osmium tetroxide, working in a fume hood and wearing disposable plastic gloves is recommended.

With chemical fixation post-mortem changes are characteristic of material that has been killed but not fixed. These changes include loss of endoplasmic reticulum and bubbling or vesicularization of membranes, with mitochondria surviving up to 24 hr (Ito, 1962). It is helpful to note that the most common problem in tissue preparation is not the post-mortem changes that might take place, but the damage resulting from poor handling.

2 FIXATION

2.3 Chemical Fixatives

Several good fixatives have come into general use for the preparation of specimens for electron microscopy (osmium tetroxide, OsO_4; the aldehydes: formaldehyde, HCHO; glutaraldehyde, $CHO(CH_2)_3CHO$; acrylic aldehyde, CH_2CHCHO; and potassium permanganate, $KMnO_4$).

Regardless of the chemical fixative used, certain features of the cell substructure may be used to determine the quality of fixation. Mitochondria should not be swollen or empty looking and their membranes should be smooth and roughly parallel. The nucleus should appear finely granular with the nuclear envelope having both membranes continuous and parallel. The plasmalemma should be continuous around the cell and smooth. Golgi bodies should consist of intact smooth membranes with perhaps vesicles at the edges. The endoplasmic reticulum should again be continuous with parallel membranes. The tonoplast, like the plasmalemma, should be continuous without unusual wrinkled regions. Usually, any clumping of the granular matrix of the cytoplasm or blebbing of cytomembranes indicates improper fixation.

In devising formulae for fixatives, the factors considered include pH, tonicity, and the quantity of divalent ions. Specific recipes for fixatives and their buffers are given later in this section, but first general factors affecting the fixation process are discussed.

Buffer Unbuffered fixatives give variable results. Therefore solutions of various salts have been used to prevent wide ranges in pH in fixatives. These buffering solutions also prevent a possible acidic wave of injury as the fixative penetrates the cell. The buffers used include the veronal-acetate buffer of Michaelis (see Palade (1952) and Luft (1956)), the chromate buffer of Dalton (1955), the phosphate buffer of Millonig (1961a), and the cacodylate buffer of Sabatini *et al.* (1962).

Most tissues are fixed near the optimum physiologic pH values of 7.2 to 7.5 for animals and a pH value of 6.8 for plant material. Hence the need of a buffer as a stabilizer to prevent changes in pH. However, some exceptions are the nuclear regions of bacteria which yield better fixations at a value of pH 6 and highly dehydrated tissues and marine plant material which show a better cytoplasmic organization at pH values of 8 to 8.5.

Tonicity An effort is usually made to have the fixative isotonic to the cell fluids of the specimen. However, whether tonicity is a factor and what level of tonicity to use is best determined by trial and error with the particular specimen being studied. Tonicity of a fixative can be adjusted with addition of sucrose or various balanced salt solutions.

Tonicity can be estimated by calculation from molarity or it can be measured with greater accuracy directly by determination of its freezing point. Marine specimens may be fixed in normal or artificial sea water to insure the proper tonicity.

Temperature Some fixatives require low temperatures to prevent over-fixation (for example permanganates). Over-fixation may result in oxidation of cytomembranes and leaching out of various cell components. On the other hand, slow penetrating fixatives such as osmium may allow cell autolysis. The fixative solution can be precooled in an ice bath to control the rate of fixation, and the tissue placed in the cooled fixative. Fixation can then either be continued at the low temperature (e.g., 4° C) or allowed to come to room temperature.

Duration of the fixation process The actual time of the fixation process is dependent on temperature, the fixative, and the tissue. It is wise to try a series of fixation periods. The series might run for 15 min, 1, 2, 4, or 8 hr.

Rinsing Rinsing in the appropriate buffer should be carried out with a minimum of three to four changes in the cold rinse. Rinses should be about 15 min apart whereas the final rinse can be overnight at the 4°–8° C. Rinses may contain "post staining" substances such as ruthenium red (used after fixation with osmium tetroxide) or uranyl acetate in concentrations of 0.5 to 1%. These stains will be discussed further in chapter 10.

Presence of divalent ions Some investigators have found that divalent ions such as calcium or magnesium are helpful in binding specific cell structures such as the DNA material of cells and the middle lamellae of plant cell walls.

Types of chemical fixatives The cell structure resulting after a chemical fixation is often distinctive for that fixative. From an electron micrograph of a sectioned eukaryotic cell, the fixative can, in most cases, be correctly identified. Figures 2-3 through 2-8 compare sections of pituitary gland tissues of a fish fixed, respectively, in osmium tetroxide, potassium permanganate, glutaraldehyde, and acrylic aldehyde, or some combination of these fixatives. Note that the cell structure appears somewhat distinct for each fixative because of the reactions of that fixative with various components of the cell (Table 2-1).

2   FIXATION

Figure 2-4 Fixation with 0.6% potassium permanganate with Luft's veronal-acetate buffer. All membranes are well preserved and quite pronounced. The background material including the secretory droplets are less distinct. The cristae of the mitochondria appear collapsed. The endoplasmic reticulum appears to be distorted ($\times$ 9200).

Figure 2-5 Fixation in 6% glutaraldehyde followed by a postfixation in 1% osmium tetroxide, both in Millonig's phosphate buffer. The endoplasmic reticulum and nuclear envelope appear swollen, perhaps caused by tonicity differences between the cell and buffer. Fixation is typical for this double fixative with good preservation of background (cytoplasmic) proteins and phospholipid membrane systems ($\times$ 9200).

Figure 2-6 Fixation in 6% glutaraldehyde followed by 1% potassium permanganate in Holt's cacodylate buffer. The nuclear envelope and mitochondrial envelope are well fixed although some swelling appears to have occurred (perhaps from the permanganate postfix). The dense black granule on the central nucleus is a crystal of lead carbonate ($\times$ 9200).

Figure 2-7 Fixation in 3% acrylic aldehyde, 3% glutaraldehyde followed by 0.6% potassium permanganate in Holt's cacodylate buffer. The cytoplasmic fixation appears coarser in granularity than that fixed with glutaraldehyde alone. The secretory bodies are well fixed and the membranes appear somewhat granular ($\times$ 9200).

CHEMICAL FIXATIVES 2.3

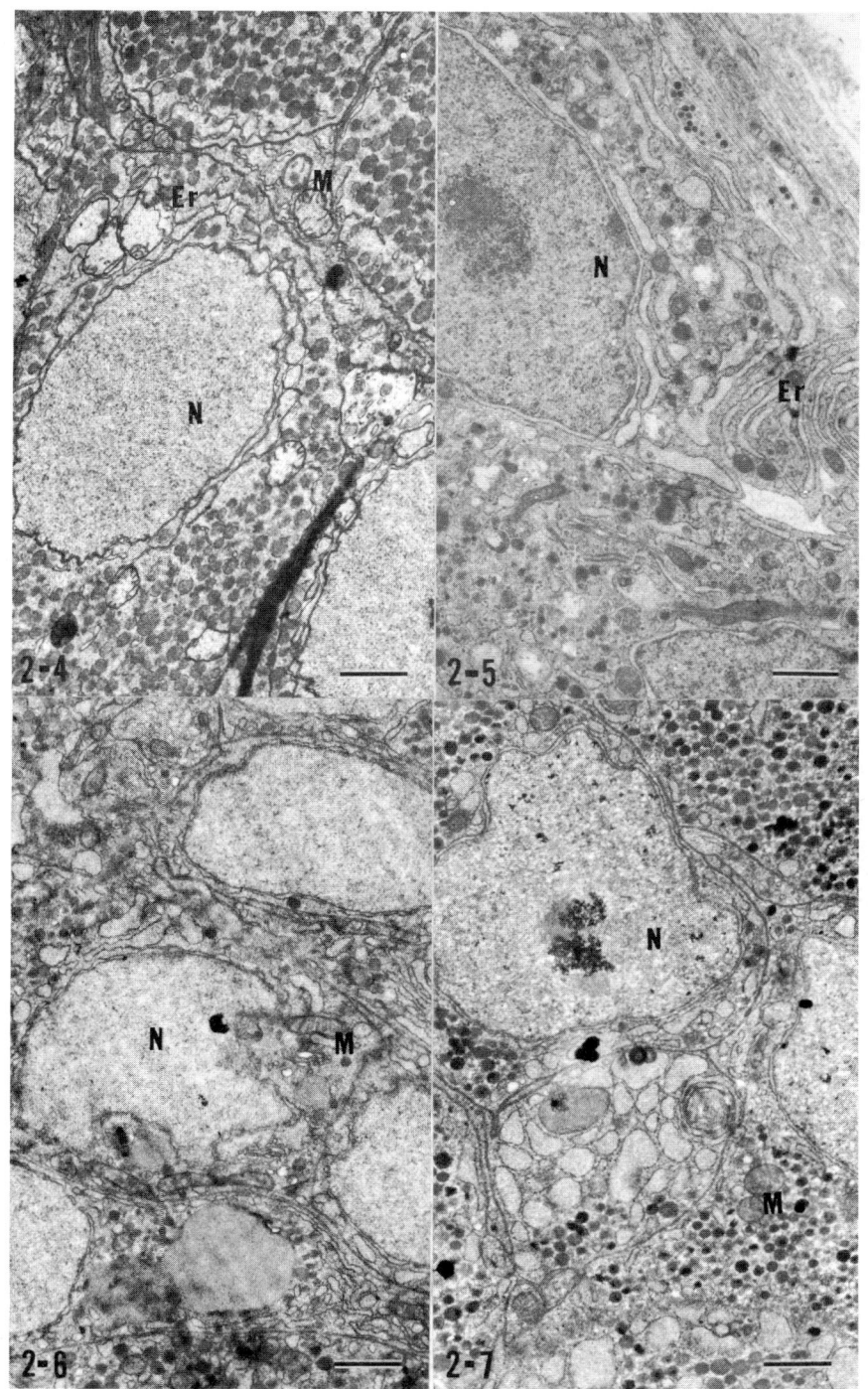

2 FIXATION

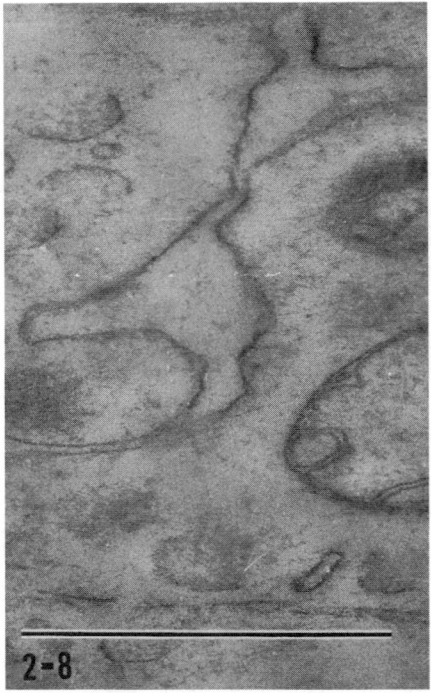

Figure 2-8 Fixation in 0.6% potassium permanganate in Luft's veronal-acetate buffer. This preparation was overfixed for 1 hr. Compare the loss of cytoplasmic detail in this micrograph with that of Fig. 2-4. The membranes are badly distorted ($\times$ 49,300).

The three most common fixatives used in the chemical fixation process are osmium tetroxide, the aldehydes (formaldehyde, glutaraldehyde, and acrylic aldehyde), and the permanganates including potassium and barium permanganate. A few other more specialized compounds are uranyl acetate, the chromium salts, the mercury salts, and phosphotungstic acid. For comparison, Table 2-1 is a summary of the reactions of cell constituents to various fixatives. The three main fixatives are now discussed.

Osmium Tetroxide (OsO$_4$)

Osmium tetroxide (osmic acid, OsO$_4$) was known to provide superior preservation of cytoplasmic detail (Strangeways and Canti (1927)). However, it was little used in light microscopy because of its high cost, poor penetration, and its interference

TABLE 2-1 General reactions of fixatives with cell constituents.

Fixative	Polysaccharide	Nucleic Acid[a]	Protein	Phospholipids	Fats[b]
Osmium tetroxide (OsO_4)[c]	No known reaction (although starch and glycogen seem affected)	Partially fixed (probably because of histones)	Partially fixed	Easily fixed	Unsaturated fats easily fixed
Aldehydes (HCHO)[c,d]	No reaction	Partially fixed (probably because of histones)	Easily fixed	No reaction	No reaction
Permanganates (MnO_4)[c]	Partial fixation, destruction occurs if overfixed	No reaction, destruction occurs if overfixed	Easily fixed but then destroyed	Easily fixed	No reaction
Chromium salts[f]	No reaction, destruction occurs if overfixed	No reaction	Easily fixed	No reaction	No reaction
Mercuric chloride ($HgCl_2$)[g]	No reaction	Coagulation occurs	Coagulation occurs	No reaction	No reaction
Uranyl salts[h]	Some reaction with galacturonic acids	Good fixation	Easily fixed	No reaction	No reaction
Phosphotungstic acid[i]	No reaction	No reaction	Easily fixed	No reaction	No reaction

[a] In general, most fixatives do not react with nucleic acids (except chromyl salts) but do react with associated proteins.
[b] Saturated fats are leached out because of no fixation.
[c] Baker, J. R. (1965), "The Fine Structure Produced in Cells by Fixatives," *J. Roy. Microsc. Soc.* **48**, 115–131.
[d] Holt, S. J., and R. M. Hicks (1961b), "Studies on Formalin Fixatives for Electron Microscopy and Cytochemical Staining Purposes," *J. Biophys. Biochem. Cytol.* **11**, 31–45.
[e] Luft, J. H. (1956), "Permanganate—A New Fixative for Electron Microscopy," *J. Biophys. Biochem. Cytol.* **2**, 799.
[f] Dalton, A. J. (1955), A "Chrome-Osmium Fixative for Electron Microscopy," *Anat. Record.* **121**, 281.
[g] Baker, J. R., and Luke, B. M. (1963), "The Fine Structure Produced in Cells by Primary Fixatives. Mercuric Chloride," *Quart. J. Micr. Sci.* **104**, 101–106.
[h] Marinozzi, V., and A. Gautierr (1962), "Fixations et Colorations. Étude des Affinités des Composonts Nucléoproténiques pour l'Hydroxyde de Plomb et l'Acetate d'Uranyle," *J. Ultrastructure Res.* **7**, 436–451.
[i] Wohlforth-Bottermann, K. E. (1956), "Die Entstehung, die Vermehrdung und die Abscheidung geformter Sekrete der Mitochondrien von *Paramecium*," in F. S. Sjöstrand and J. Rhodin. Eds., *Electron Microscopy Proc. Stockholm Conference 1956* (Almquist and Wiksell, Uppsala), 137–139.

2 FIXATION

with certain staining methods. In the 1950s a number of laboratories experimented with the use of this fixative in electron microscopy. Palade (1952) was one of the first to obtain reproducible results with this fixative by using the veronal-acetate buffer of Michaelis (see also Luft (1956)). There are a number of buffers that have been shown experimentally to work well with osmium tetroxide.

Table 2-1 shows the reactions of osmium tetroxide with the basic cell constituents. In general, the ground cytoplasm of a eukaryotic cell fixed with osmium appears granular because the cytoplasmic protein is preserved (Fig. 2-3). Cytomembranes are usually sharp and the nucleus appears somewhat fibrillar or coarsely granular. The effect on the nucleus may be the result of condensation of the histones in DNA and thus should be considered an artifact. Unsaturated fats, usually in the form of droplets found in chloroplasts and fat vacuoles are well preserved with osmium. Cell wall material, starch grains, and glycogen appear to be nonreactive to osmium, consequently their presence must be determined by subsequent treatment of the tissue (see Fig. 2-3 for an example of tissues fixed with osmium tetroxide).

Fixation procedures Since osmium tetroxide fumes are harmful to both the eyes and respiratory tract, work with these solutions should be carried out in a fume hood. This fixative is a very reactive reagent; therefore, glassware must be clean and the solutions uncontaminated. Glass stoppered bottles should be used and the solutions should be kept in the refrigerator, preferably in a dissector jar.

Osmium tetroxide is normally supplied as crystals in ampules, which should be washed thoroughly to remove all traces of the label and organic glue. The ampule is broken with a file and is placed in the buffer solution or water. To prevent loss of loose crystals of osmium tetroxide, prior to breaking the ampule, melt the crystals by placing the ampule under warm tap water and then allow the liquid to solidify into one lump. Osmium crystals dissolve slowly at room temperature; dissolution may take two to three days. Dissolution can be hastened by exposing the osmium crystals to ultrasonic vibration or by heating the solution to 60° C. Care must be taken when heating the osmium solutions. Final concentration of the osmium solution for fixation should be in the range of 1–2% with slightly better results using 2%.

Since osmium tetroxide is a slow penetrant, the tissue is usually cut up or minced in a small drop of this fixative to yield sections

CHEMICAL FIXATIVES 2.3

less than 1 mm in thickness. These sections may be transferred with the flat end of a toothpick to a vial containing fresh fixative.

The duration of the fixation process varies according to the fixative, the specimen, and the temperature. For osmium, at 4° C, a fixation time of 15 min to 2 hr is usually sufficient in a 1% solution. As osmium tetroxide is reduced the specimen will turn black with osmium particles. This blackening of the tissue usually indicates that the tissue is fixed. Over-fixation results in the leaching out of the material and the vesiculization (bubbling) of the cytomembranes. A series of fixation periods should be tried to determine the optimum fixation time for a given tissue.

Most fixations using osmium described in the literature were carried out from 4° C to room temperature. Pease (1964) found subtle qualitative differences depending on the temperature and obtained the best results when fixation was begun at 0° C and then allowed to reach room temperature. This is also the author's standard procedure.

Following fixation in osmium, it is recommended that at least four rinses in fresh buffer without fixative be made. This removes any free osmium (not reduced) which otherwise would precipitate in the cells forming black (electron dense) granules. Each rinse should be cold and last about 15 min. The specimen can then be dehydrated or, if necessary, stored overnight for dehydration the next day. Storage in the buffer for longer than 1 day is not recommended because of the possibility of leaching constituents from the cell.

Some investigators "post fix" the tissues by placing them in formalin (a 10% solution) which has been stored over calcium carbonate. This step supposedly toughens the tissue and helps to prevent damage during dehydration and embedding, and may contribute to better preservation. The usual time in formalin is 1 min to 2 hr although the tissue may be left overnight.

Buffer formulae for osmium tetroxide The formulae listed below are organized as follows: title, any recommended uses on the solution, stock solutions to be available, fixative constituents, duration and temperature to be used in fixation, and other pertinent information (pH adjustment, tonicity).

(1) Palade (1952), Veronal-acetate buffered osmium (1%) This is the original osmium tetroxide buffer solution and is still in wide use for standard fixation of both animal and plant tissue.

2 FIXATION

Stock Solutions:
  a. Buffer stock solution (0.28M)

sodium veronal (sodium barbital)	14.7 g
sodium acetate (hydrated, crystalline form)	9.7 g
or anhydrous, powdered form	(5.7 g)
water	500.0 ml

  b. 0.1N Hydrochloric acid (HCl)

concentrated HCl (36%, 11.6M)	0.86 ml
water	100.0 ml

  c. Stock osmium tetroxide (2%)

crystalline OsO_4	2.0 g
water	100.0 ml

Note: Each of these solutions is stable and can be stored in brown bottles at room temperature or in a refrigerator. The osmium tetroxide stock solution will discolor (blacken) if contaminated, and a fresh solution should be made up when this occurs.

Fixative Constituents:

Buffer stock solution	2 vol
0.1N HCl, titrate to desired pH (*ca.* 7.2)	*ca.* 2 vol
Distilled water to make 5 vol	*ca.* 1 vol
2% OsO_4	5 vol
	10 vol

Note: First, adjust the pH of the buffer stock solution to the desired value (usually 7.2 to 7.4) by adding approximately equal amounts (2 vol each) of the 0.1N HCl and the buffer stock solution. The last portion of the HCl should be added slowly (titrated), while the pH is constantly checked by a pH meter. When the pH is adjusted, enough distilled water should be added (about 1 vol) to bring the total volume up to $2\frac{1}{2}$ times the original volume of the buffer stock. This "neutralized buffer stock" is mixed with an equal volume of the 2% OsO_4 stock solution to produce the final fixative, a 1% OsO_4 solution. Since "neutralized buffer solution" is unstable and will crystallize if left standing, it must be mixed with the 2% OsO_4 shortly after the HCl solution has been added.

Duration and Temperature: Palade recommended rather short periods of exposure to the fixative to prevent leaching out of proteinaceous material, 30 to 60 min is usually sufficient. Fixations up to 2 hr do not appear deleterious. Osmium fixation appears to be effective if conducted either entirely in the cold or if allowed to rise to room temperature during fixation.

(2) Dalton (1955), Chrome-osmium fixative (1%) This fixative was recommended because:

1. At a pH of 7.2 the fixative causes cross-linkage between protein molecules.
2. It is stable with respect to its pH value and reduction of osmium.
3. There is no leaching of cell constituents, even after long periods of fixation.
4. The duration of the fixation process may be varied from 2 to 24 hr.

Although this fixative is not in common use today, it is included as a possible fixative using osmium which is a most effective preservative of cytoplasmic protein.

Stock Solutions:

a. 3.4% sodium chloride (NaCl)	3.4 g/100 ml H_2O
b. 2.5N potassium hydroxide (KOH)	14.0 g/100 ml H_2O
c. Potassium dichromate-potassium hydroxide buffer ($K_2Cr_2O_7$-KOH)	
5% $K_2Cr_2O_7$	5.0 g/100 ml H_2O
2.5N KOH	approximately 12–16 ml
Distilled water	to make 125 ml (see Note below)
d. 2% osmium tetroxide	1.0g/ 50 ml H_2O

Note: Titrate the 2.5N KOH into the $K_2Cr_2O_7$ solution until a pH of 7.4 is reached, then add sufficient distilled water to make 125 ml of completed buffer.

Fixative Constituents:

2% osmium tetroxide	2 vol
4% potassium dichromate-potassium	2 vol
hydroxide buffer	1 vol
3.4% sodium chloride	1 vol

Note: The $K_2Cr_2O_7$ and KOH mixture forms a buffer which is stable between pH of 5.6 to 7.6. However, since chrome apparently is active on proteins at a pH of 7.4, this is the recommended end point in titration.

Duration and Temperature: Dalton recommended about 2 hr for fixation. Based on previous experiments, this seems reasonable if low temperatures are used (4° C). Rinse at least four times (15 min each) in cold buffer.

(3) Millonig (1961b), Phosphate buffered osmium (1%) This is the standard buffer used for osmium fixation in the author's laboratory.

2 FIXATION

Stock Solutions:
  a. 2.26% monosodium phosphate
     ($NaH_2PO_4 H_2O$) 2.26 g/100 ml H_2O
  b. 2.52% sodium hydroxide
     (NaOH) 2.52 g/100 ml H_2O
  c. 5.4% glucose 5.4 g/100 ml H_2O
  d. 1.0% calcium chloride
     ($CaCl_2$) 1.0 g/100 ml H_2O
  e. Osmium tetroxide (OsO_4) 0.5 g ampules

Note: These stock solutions are stable if kept in the cold.

Fixative Constituents:

2.26% $NaH_2PO_4 \cdot H_2O$	41.5 ml
2.52% NaOH	8.5 ml
5.4% glucose	5.0 ml
1.0% $CaCl_2$ (pipette in)	0.25 ml
OsO_4	0.5 g

Note: Mix together thoroughly the 2.26% $NaH_2PO_4 \cdot H_2O$ and the 2.52% NaOH solutions in the specified quantities. Remove 5 ml of this mixture and test the pH which should be 7.3 if all the solutions of the buffer were properly prepared. To the remaining 45 ml of this solution, add the specified amounts of glucose and 1% $CaCl_2$. To this resulting 50 ml of buffer solution add 0.5 g of OsO_4 to make a 1% osmic acid solution for fixation. As noted previously, the OsO_4 crystals will require some time to dissolve.

The addition of NaOH to $NaH_2PO_4 \cdot H_2O$ forms an isotonic disodium phosphate buffer. By varying the amounts of NaOH added, the pH of the final buffer can be easily adjusted between 5.4 and 8 without changing the tonicity. Calcium is thought to help the preservation of certain structures, e.g., nuclear structure, phospholipids, and cytomembranes. This buffer is said to be more efficient than veronal-acetate in maintaining pH and tonicity deep in the specimen because both mono- and di-sodium phosphate already exist as effective buffers in most body fluids. Furthermore, Millonig (1961b) feels the use of sodium salts avoids any possible toxic effects that might result if potassium were used.

Duration and Temperature: In most cases, this phosphate buffer with 1% OsO_4 effectively fixes small blocks of tissue within 1–2 hr. Again, it is recommended that the tissue be placed

CHEMICAL FIXATIVES   2.3

in cold fixative and then be allowed to come to room temperature during fixation.

(4) McClean and Cook (1952), Sorensen's phosphate buffer and osmium (1%)   This buffer has been employed by a number of laboratories in osmium tetroxide fixation, including the author's.

Stock Solutions:
  a. 0.066M dibasic sodium phosphate
     ($Na_2HPO_4 \cdot 2H_2O$)   11.876 g/ H_2O
  b. 0.066M monobasic potassium
     phosphate (KH_2PO_4)   9.078 g/ H_2O
  c. osmium tetroxide (OsO_4)   1 g ampules

Fixative Constituents:
  $Na_2HPO_4 \cdot 2H_2O$ 78.0 ml
  KH_2PO_4 22.0 ml
  OsO_4 1.0 g

Note: The above preparation should have a pH of 7.4. As shown in Table 2-2, a range of pH can be achieved by varying the amounts of the two solutions.

TABLE 2-2   Sorensen's phosphate buffer formula.*

pH	ml 0.066M $Na_2HPO_4 \cdot 2H_2O$	ml 0.066M KH_2PO_4
6.0	1.4	8.6
6.2	2.0	8.0
6.4	3.0	7.0
6.6	4.0	6.0
6.8	5.0	5.0
7.0	6.1	3.9
7.2	7.0	3.0
7.4	7.8	2.2
7.6	8.5	1.5
7.8	9.1	0.9

* Note: See text for formulae to make up the two 0.066M solutions.

Duration and Temperature: This buffer, as with Millonig's phosphate buffer (previous formula), seems to be effective within 2 hr on small blocks of tissue. Follow the outlined procedure for starting fixation at 4° C, and allow the fixative and tissue to come to room temperature.

(5) Kellenberger et al. (1958), Veronal-acetate buffered osmium (1%)   This buffer, with a pH of 6.1, is considered useful for fixation of bacteria, cell suspensions, and homogenates.

2 FIXATION

Stock Solutions:
   a. Veronal-acetate-sodium chloride buffer

sodium veronal (sodium barbital)	14.7 g
sodium acetate ($NaC_2H_3O_2 \cdot 3H_2O$)	9.7 g
sodium chloride (NaCl)	17.0 g
distilled water to make	500 ml solution
b. 1M $CaCl_2$	110.99 g/l H_2O
c. 0.1N HCl	0.86 ml conc HCl/100 ml H_2O
d. osmium tetroxide (OsO_4)	0.5 g ampules

Fixative Constituents:

Stock veronal-acetate-sodium chloride buffer	10.0 ml
0.1N HCl	14.0 ml
1M $CaCl_2$	0.5 ml
distilled water	26.0 ml
osmium tetroxide	0.5 g

Duration and Temperature: Kellenberger and coworkers recommend that 30 ml of culture material be mixed with 3 ml of this fixative at room temperature, and that this mixture be centrifuged immediately for 5 min at about 4000 rpm (1800 g). The prefixed material is concentrated into a pellet which is resuspended in 1 ml of fresh fixative and 0.1 ml of the growth medium and then left overnight at room temperature. After about 16 hr of fixation the suspension is diluted with 8 ml of only the buffer solution and recentrifuged. The resulting pellet is resuspended for a second time in 1 ml of 2% warm agar. After cooling, the agar block is soaked in 0.5% uranyl acetate (aqueous) for 2 hr at room temperature, rinsed, and dehydrated.

Aldehydes (HCHO)

Three aldehydes commonly used in fixations for electron microscopy are formaldehyde (HCHO), glutaraldehyde ($CHO(CH_2)_3$-CHO), and acrylic aldehyde ($CH_2 \cdot CHCHO$). These fixatives are especially useful in histochemical studies since formaldehyde and glutaraldehyde do not block various enzymatic functions. Hence, histochemical localization of enzymes can be performed on sectioned material.

The most common feature of aldehyde fixed tissue is the rich and granular micrograph image of the cell cytoplasm which is due to the preservation of proteinaceous material (Table 2-1).

Since an aldehyde yields an overall "bland image" of the cell, there are no electron dense metals in the fixative. Postfixation is usually desired. In fact, if there is no postfixation, the cytomembranes in the micrograph of an aldehyde fixed cell will appear as negative images of those seen in a micrograph of an osmium fixed cell. Osmium tetroxide or potassium permanganate should be used in the postfixation process since they contain electron dense metals (see Figs. 2-5, 2-6, and 2-7 for examples of aldehyde fixed tissue).

Fixation procedures   Reasonably large pieces of tissue—2 cm blocks—can be fixed by the aldehydes due to their rapid penetration. Furthermore, the duration of the fixation process and the concentration of the fixative used do not appear to affect the resulting micrographs. Usually the tissue is soaked for 15 min to 2 hr in 5 to 10% solutions of aldehyde, but overnight fixation apparently does no harm. Temperature is not as critical in aldehyde fixation. Room temperature is commonly used, especially if microtubules are to be preserved.

The aldehydes tend to break down when stored to form acidic products such as formic acid and acrylic acid. Stock solutions of the aldehydes are best kept refrigerated in glass-stoppered bottles with activated charcoal. If this is done, the pH value should reach almost 7. To raise the pH of the aldehyde, add insoluble barium carbonate and allow to settle or distill the fixative.

As mentioned above, because of the bland images resulting from a straight aldehyde fixation, a heavy metal postfixative such as osmium tetroxide or potassium permanganate is usually employed. In most cases the same buffer used with the aldehyde should be used with the postfixative. Duration of postfixation is usually longer than the primary fixation. The initial aldehyde fixation seems to prevent the postfixative from penetrating at its normal rate. Prior to or after postfixation, an exposure to 0.5% uranyl acetate in the appropriate buffer for a few hours to overnight may be done. This enhances the preservation of nucleic acid in the nucleus, mitochondria, and chloroplasts.

Since the aldehydes reduce osmium if they are used as a postfixative, it is important to adequately rinse the specimen before proceeding with postfixation. Aldehydes tend to coat the tissue surface and at least three short rinses in the buffer solution should be made, each of at least 15 min duration followed by a final soak in the buffer (cold) for 1 hr to overnight.

Buffer formulae for aldehydes   Each formula is organized as follows: title, stock solutions, fixative constituents, duration and

2 FIXATION

temperature, and postfixation procedure. Unless noted otherwise the aldehyde in discussion can be used with other buffers given for aldehydes.

(1) Pease (1964), Phosphate buffered formaldehyde (10%)

Stock Solutions:
- a. 2.26% monosodium (monobasic) sodium phosphate ($NaH_2PO_4 \cdot H_2O$) — 2.26 g/100 ml H_2O
- b. 2.52% sodium hydroxide (NaOH) — 2.52 g/100 ml H_2O
- c. 40% formaldehyde (10% formalin) — 40.0 g /100 ml H_2O

Note: The use of formalin (reagent grade) is not recommended by Pease because of the presence of methanol in the solution. Consequently paraformaldehyde, a simple polymer of formaldehyde is used as follows in preparing the formaldehyde solution. Dissolve 40 g of paraformaldehyde at 60° C in 100 ml water by adding drops of 0.1N NaOH until the solution clears (at about a pH of 7.2). A slight milkiness may persist until the solution is diluted with a buffer.

Fixative Constituents:

2.26% monosodium phosphate	62.2 ml
2.52% sodium hydroxide	12.8 ml
40% formaldehyde	25.0 ml

Duration and Temperature: Because formaldehyde can penetrate rapidly, even in large blocks (cm³) of tissue, the duration of fixation can be 30 min to 4 hr in the cold. Pease recommends that the fixative be cold and then the process be allowed to come to room temperature.

Postfixation: The 1% osmium tetroxide is recommended after thorough rinsing. The blocks should be cut down to 1 mm cubes, if possible, to insure thorough penetration of osmium tetroxide. The temperature may be either 4° C or room temperature.

(2) Sabatini et al. (1962), Cacodylate buffered glutaraldehyde (5%)

Glutaraldehyde is a 5-carbon molecule with an aldehyde group at each end and an odor not unlike that of cider. It is a good general purpose fixative and is especially useful in histochemistry because at least some cellular enzymatic activities are preserved (oxidase, alkaline phosphatase, acid phosphatase (Sabatini et al. (1963)). The use of cacodylate buffer ($Na(CH_3)_2 AsO_2 \cdot 3H_2O$) in histochemical studies avoids the

CHEMICAL FIXATIVES 2.3

presence of extraneous phosphates which may act as toxins. The buffer also has a good range within which the pH is stable (see Table 2-3). Amino acid-containing buffers are not used with glutaraldehyde since they can be expected to react with the aldehyde.

TABLE 2-3 The pH values of $0.5M$ sodium cacodylate buffer. A $0.2M$ solution of sodium cacodylate is made up (4.28 g/100 ml H_2O); sucrose can be added if desired.

To each 100 ml of the $0.2M$ sodium cacodylate, add the following amounts of $0.2M$ HCl (to 0.729 ml conc HCl (37.4%)), add water to a volume of 37.4 ml to obtain $0.2M$ HCl) to obtain the desired pH. Finally, dilute the solution with distilled water to reach a total volume of 400 ml of $0.05M$ sodium cacodylate.

$0.2M$ HCl	pH
5.4	7.4
8.4	7.2
12.6	7.0
18.6	6.8
26.6	6.6
36.6	6.4
47.6	6.2
59.2	6.0
69.6	5.8
78.4	5.6
86.0	5.4
90.0	5.2
94.0	5.0

Stock Solutions:
  a. Fixing buffer
     $0.2M$ sodium cacodylate
     $(Na(CH_3)_2AsO_4 \cdot 3H_2O)$ (in
     anhydrous form use
     3.2 g/100 ml H_2O)   4.28 g/100 ml H_2O
  b. Rinsing buffer
     $0.1M$ sodium cacodylate (in
     anhydrous form use
     1.6 g/100 ml H_2O)   2.14 g/100 ml H_2O
     0.1N HCl   approximately 4 to $8\frac{1}{2}$ ml/100 ml H_2O
     Sucrose   7.3 g/100 ml H_2O
  c. $0.2M$ HCl   7.29 ml conc HCl/374 ml H_2O
  d. 25% Commercial glutaraldehyde

2 FIXATION

Note: The pH of the rinsing buffer is adjusted to 7.4 by adding 0.1N HCl, and then the sucrose is added for tonicity. The 25% glutaraldehyde should be stored over activated charcoal to prevent increase in acidity.

Fixative Constituents:
25% glutaraldehyde	20 ml
0.2M sodium cacodylate (titrate to desired pH value) not more than	80 ml
Add water to make	100 ml

Note: Sabatini and coworkers recommend a pH of 7.4 and their method will result in a molarity of about $0.1M$ for the fixing solution. An alternative method, using $0.05M$ sodium cacodylate can be used if a much lower molarity is desired (Table 2-3). With this table, the pH value of the buffer is first adjusted to the desired level and then the glutaraldehyde is added to reach the desired concentration of the fixative. For a 5% solution take 5 ml of 25% glutaraldehyde and add 20 ml of $0.05M$ sodium cacodylate buffer (Table 2-3).

Duration and Temperature: Fixation for 15 min to 2 hr in a cold fixative is usually sufficient. The fixative can be allowed to come to room temperature during fixation.

Postfixation: Both 1% osmium and 0.6% potassium permanganate have been used with success following the primary fixation with glutaraldehyde in cacodylate buffer. Thorough rinsing in the buffer is necessary to remove any glutaraldehyde before postfixation in 1% osmium. Length of postfixation should be 1 to 2 hr for osmium and about 1 hr for permanganate. The same buffer can be used for postfixation and the final rinsing before dehydration.

(3) Holt and Hicks (1961b), s-Collidine buffered glutaraldehyde (5%) This fixative is apparently quite effective on both plant and animal material, causing little damage to cell membranes, even when used with marine plant tissue. However, s-collidine does leach out cytoplasmic proteins.

Stock Solutions:
a. Stock s-collidine (trimethylpyridine)	2.67 ml
b. Commercial grade 25% glutaraldehyde	

Note: Add 2.67 ml of s-collidine to 50 ml of distilled water. Then titrate 0.1N HCl into the above solution until a pH of 7.2 is reached. Finally add sufficient water to make 100 ml of solution.

CHEMICAL FIXATIVES 2.3

Fixative Constituents:
    25% glutaraldehyde   20 ml
    stock s-collidine buffer solution
      (trimethylpyridine)   80 ml

Duration and Temperature: Usually 15 min to 2 hr is sufficient; however, effective fixation may require up to 6 hr. Start fixation cold (4° C) and allow the process to attain room temperature. If histochemical studies are to be done, a shorter fixation (10 to 30 min) and continuous low temperature (4° C) are recommended.

Postfixation: After thorough rinsing, postfixation can be best carried out with 1% osmium or 0.6% potassium permanganate (Fig. 2-6). Postfix for 2 hr at room temperature.

(4) Millonig (1961b), Phosphate buffered glutaraldehyde (10%) This fixative has yielded excellent results for plant tissues and gives an expanded appearance to cell organelles (mitochondria and chloroplasts) in the micrographs. However, in some cases cell shrinkage does occur, perhaps because of the hypertonicity of glutaraldehyde, and consequently lower fixative concentrations are in order. Before using this glutaraldehyde fixative, be certain that all acidic by-products are removed from the stock aldehyde solution.

Stock Solutions:
    a. Millonig phosphate buffer (see previous section on osmium fixation)
    b. 25% Commercial glutaraldehyde

Fixative Constituents:
    25% glutaraldehyde   40 ml
    stock phosphate buffer   60 ml

Duration and Temperature: Usually 2 to 6 hr will be adequate. However, to prevent blebbing of cytomembranes, try some material for about 30 min and start all tissue at 4° C allowing the process to come to room temperature during fixation.

Postfixation: Apparently the most effective postfixative is 1% osmium tetroxide in the same buffer and cold (4° C) for about 2 hr (Fig. 2-5). Insure, by thorough rinsing, that all the glutaraldehyde has been removed before postfixation.

(5) Gibbs (1962), Veronal-acetate-salt buffered glutaraldehyde (5%) This buffer incorporated Sjöstrand's (1967) physiological buffer as a salt solution to match the molarity of sea water (0.85M). The higher molarity makes the buffer especially

2 FIXATION

useful in fixation of marine organisms such as algae and fungi. In some cases the micrographs of plant material fixed with this buffer show strong compression of cell organelles; that is, chloroplasts and mitochondria appear flattened, with lamellae closely appressed.

Stock Solutions:
    a. Sodium Veronal-Acetate Buffer (see Palade's veronal-acetate buffer for osmium)
    b. 0.1N HCl
        concentrated HCl (36%, 11.6M)   0.86 ml/100 ml H_2O
    c. Salt Solution

sodium chloride	40.25 g
potassium chloride	2.19 g
calcium chloride	0.90 g
distilled water	to make 500 ml

    d. 25% Commercial glutaraldehyde

Fixative Constituents:

sodium veronal-acetate buffer	20.0 ml
salt solution	29.2 ml
0.1N HCl	approximately 22.0 ml
distilled water	to make 100 ml solution

Note: To 80 ml of the above described buffer-salt solution add 20 ml of 25% glutaraldehyde for a 5% glutaraldehyde fixative.

Duration and Temperature: As with most aldehydes, fixation is usually completed within 30 min to 6 hr at room temperature although the specimen can be left in cold fixative (4° C) and allowed to come to room temperature during fixation.

Postfixation: Either 1% osmium tetroxide or 0.6% $KMnO_4$ in the same buffer as glutaraldehyde are recommended. After thorough rinsing, the tissue should be postfixed in the cold for 1 to 2 hr. Note that since osmium tetroxide will react with this buffer-salt, prepare only small amounts for postfixation.

(6) Ramus (1969), Cacodylate-sucrose buffered glutaraldehyde (5%) This schedule was used for preservation of a marine red alga, a difficult group of plants to fix.

Stock Solutions:
    a. 0.1M sodium cacodylate   2.14 g/100 ml H_2O
        sucrose (0.25M)   8.558 g
    b. 25% Commercial glutaraldehyde

CHEMICAL FIXATIVES 2.3

Fixative Constituents:

25% glutaraldehyde	20 ml
0.1M cacodylate buffer with 0.25M sucrose	80 ml

Note: The addition of 0.25M sucrose in the primary fixative glutaraldehyde, is for the purpose of adjusting the osmotic pressure of the buffer to that of sea water. This osmotic pressure can be varied by changing the concentration of sucrose from 0.1M to 0.5M.

Duration and Temperature: Ramus recommends 3 hr for aldehyde fixation carried on at 1° C.

Postfixation: The 2% osmium tetroxide is used in the same buffer without sucrose for 3 hr at room temperature. The material is first rinsed in a buffered series of sucrose concentrations (0.25M, 0.15M, and 0.05M sucrose). Following primary fixation, the final cacodylate buffer should be without sucrose.

(7) Luft (1959), Cacodylate buffered acrylic aldehyde (5%)
Acrylic aldehyde (also called acrolein, from which tear gas is made) is successful as a primary fixative because it is a very rapid and aggressive fixative. Use this fixative only in a fume hood. The cytoplasmic structure, as seen in the micrographs of material fixed with acrolein, is somewhat denser and more coarse than with glutaraldehyde fixation (Fig. 2-7).

Stock Solutions:
  a. Fixing Buffer, 0.2M Sodium Cacodylate (see formula from glutaraldehyde fixative)
  b. Rinsing Buffer, 0.1M Sodium Cacodylate (see formula from glutaraldehyde fixative)
  c. 100% Commercial acrylic aldehyde

Fixative Constituents:

acrylic aldehyde	5 ml
0.2M sodium cacodylate	95 ml

Note: This fixative can be stored for only short periods of time because it will precipitate the buffer salts. Another buffer acceptable for this fixative is Sorensen's phosphate buffer (see osmium fixative).

Duration and Temperature: Primary fixation in acrylic aldehyde is usually effective at room temperature within 2 to 3 hr. Place the tissue in cold fixative (4° C) and leave in refrigerator for about $\frac{1}{2}$ hr. Then, allow fixation to come to room temperature.

2 FIXATION

Postfixation: After thorough rinsing, the tissue can be postfixed in 1% osmium tetroxide in the same buffer (final solution should be $0.1M$ sodium cacodylate) for 2 hr at room temperature (starting with cold fixative).

(8) Karnovsky (1965), Cacodylate buffered glutaraldehyde-formaldehyde (3% each) Since 1964, a number of publications have appeared which have described combinations of aldehydes as fixatives. Probably the most popular combination is that of glutaraldehyde and formaldehyde buffered in sodium cacodylate.

Stock Solutions:
    a. Formaldehyde Solution (25%)
       paraformaldehyde   25 g/100 ml H_2O
    b. 25% Commercial glutaraldehyde
    c. $0.2M$ Sodium Cacodylate (see formula under osmium tetroxide)

Note: The formaldehyde solution is prepared as follows: Dissolve 25 g of paraformaldehyde in 100 ml of distilled water by heating the water to 60° C and adding about 12 drops of 1N NaOH while stirring constantly. The solution should be clear with addition of NaOH.

Fixative Constituents:

25% formaldehyde solution	3 ml
25% glutaraldehyde solution	3 ml
$0.2M$ cacodylate buffer	19 ml
$CaCl_2$ anhydrous	25 mg

Duration and Temperature: Fixation of 2 to 4 mm-thick specimens should be done at room temperature for 20 to 30 min, and then the specimens should be diced into small 1-mm blocks and fixed for a total of 2 to 5 hr.

Postfixation: The blocks are rinsed for 3 to 12 hr in cold $0.1M$ buffer and then postfixed in a 1–2% osmium tetroxide buffered in s-collidine for 2 hr at 1° C, (see osmium fixation for s-collidine buffer formula).

(9) Dawes (1969), Phosphate buffered glutaraldehyde-acrylic aldehyde (3% each) This fixative gave excellent micrographs of fungal cytoplasm. The acrylic aldehyde portion of the primary fixative appears to induce a slightly more granular appearing cytoplasm and denser cytomembranes.

Stock Solutions:
    a. 25% or 50% Commercial glutaraldehyde
    b. 100% Commercial acrylic aldehyde
    c. Millonig's Phosphate Buffer (see formulae under osmium fixation)

Fixative Constituents:

25% glutaraldehyde	12 ml
100% acrylic aldehyde	3 ml
Millonig's phosphate buffer	85 ml

Duration and Temperature: Fixation can proceed for 3 to 12 hr starting in the cold and being allowed to continue at room temperature. Usually 1–3 hr is effective and sufficient.

Postfixation: After at least three to four rinses, postfixation can follow in 2% osmium tetroxide in Millonig's buffer for 2 hr or 0.6% $KMnO_4$ in Luft's veronal-acetate buffer for 30 min to 1 hr. Postfixation should be carried on at 1°–4° C.

Permanganates (MnO_4)

Luft (1956) introduced potassium permanganate as a fixative for use in electron microscopy. The fixative was found to give excellent results with plant tissues, causing the cytomembranes to stand out in sharp relief against the rather vacant background of the ground cytoplasm. The image of a cell properly fixed with permanganate consists of sharply defined cytomembranes and an empty (protein-lacking) cytoplasm (Table 2-1). Permanganate rapidly fixes the protein of the cytoplasm, as well as the ribosomes, but because of its strong oxidative powers it removes the cytoplasmic protein. Glycogen and starch granules tend to be preserved while lipid material is lost in dehydration (Fig. 2-4).

Fixation procedures The time of fixation is critical because of the rapid and strong oxidative powers of permanganate. Furthermore, the temperature at which the fixation process takes place should be kept around 5° to 10° C to allow better control of the fixative as well as to prevent cell autolysis. Leaching out and swelling of cytomembrane material occurs if the duration of fixation is too long (Fig. 2-8), while too short a time results in membrane swelling. Experience has shown that a 1-mm block of tissue fixed in permanganate will have an outer band of tissue overfixed (showing loss of cell organization and constituents), an inner band of underfixed tissue (showing weakly stained cytomembranes and some swelling of membranes), and an intermediate region of tissue properly fixed. To control the rate of

2 FIXATION

fixation, carry out the process in an ice bath. Also run a series of fixation times (15 min, 30 min, 1 hr, and 2 hr) to determine the proper duration of fixation.

Dehydration should be rapid with little washing after fixation in permanganate, since material so fixed tends to swell in the alcohol series. Epoxy resins are usually used for embedding since after permanganate fixation polymerization damage to methacrylate embedded tissue may be quite extensive.

Buffer formulae for permanganate   Either potassium or barium permanganate may be used in the formulae described below. However, potassium tends to be favored over barium permanganate since it produces images of cytomembranes which appear less coarse in the micrographs. Organization of the formulae is similar to those given for osmium tetroxide and the aldehydes: title, stock solutions, fixative constituents, and duration and temperature.

(1) Luft (1956), Veronal-acetate buffered permanganate (0.6%)

Stock Solutions:
   a. Sodium Veronal-Acetate Buffer (see formula of Palade for osmium tetroxide)
   b. 0.1N HCl
      concentrated HCl (36%, 11.6M)   0.86 ml/100 ml H_2O
   c. 1.2% Potassium Permanganate ($KMnO_4$)   1.2 g/100 ml H_2O

Fixative Constituents:

Sodium veronal-acetate buffer	5.0 ml
0.1N HCl	about 5.0 ml
distilled water to make 12.5 ml	about 2.5 ml
1.2% potassium permanganate	12.5 ml

Note: The 0.1N HCl should be titrated to reach the desired pH of about 7.4. The sodium veronal-acetate buffer plus 0.1N HCl and distilled water make up an unstable "neutralized buffer." This must be used rather soon or else the buffer will precipitate.

Duration and Temperature: As noted above, all permanganate fixations should be carried on in the cold (0° to 4° C) and a series of fixation times should be run (15 min through 2 hr) to determine the proper length of exposure. Rinse the fixed material in cold neutralized buffer (made up fresh) for about 15 min and then dehydrate in rapid steps (to avoid swelling of cells).

(2) *Mollenhauer (1959), Unbuffered Permanganate (5%)* In his studies of plant cells, Mollenhauer used an unbuffered potassium permanganate solution with the fixation process carried out at either room temperature or at 5° to 10° C. Although the entire notion of using such an active fixative without a strong buffer would appear to contradict all that has been stated in this chapter, the "proof is in the pudding." The author has used this very simple fixative most successfully with both fungal and algal material (Dawes (1969a, 1969b)).

(3) *Dawes and Rhamstine (1967), Veronal-acetate-salt buffered permanganate (0.6%)* The salt solution given by Sjöstrand (1967) and used by Gibbs (1962) in her fixation of marine algae appears to work very well as a high molarity solution for permanganate fixation of marine plants. The pH values of the buffer are somewhat low when compared to that of seawater (pH 7.4 to 7.8 vs. 8.1 to 8.4).

Stock Solutions:
    a. Sodium Veronal-Acetate Buffer (see Palade's stock buffer for osmium)
    b. 0.1N HCl
       concentrated HCl (36%, 11.6M)   0.86 ml/100 ml H_2O
    c. Salt Solution of Gibbs (1962) (see formula for glutaraldehyde)
    d. Potassium Permanganate

Fixative Constituents:

veronal-acetate buffer	20 ml
salt solution of Gibbs (1962)	29.2 ml
0.1N HCl	approximately 22 ml
distilled water to make 100 ml	approximately 28.8 ml
potassium permanganate	0.6 g

Note: Titrate the 0.1N HCl into the veronal-acetate buffer to reach a pH value of 7.4 and then add the salt solution and sufficient distilled water to make 100 ml of fixative. Then add 0.6 g of potassium permanganate and stir thoroughly.

Duration and Temperature: See comments for Luft's fixative procedure (formula (1) of this section).

2 FIXATION

(4) *Dawes and Rhamstine (1967), Sea water buffered permanganate (0.6%)* The author's laboratory has obtained good results with sea water used as a buffer for potassium permanganate in fixing marine plants. The seawater will retain a pH around 7.2 to 7.4. Sea water should be Seitz-filtered twice. Potassium permanganate is mixed with the seawater until 0.6% permanganate concentration is reached. Low temperature and a series of fixation periods should be used.

2.4 Mechanical Fixation

To support and confirm the vast amount of information obtained from chemically fixed tissue and to obtain new information on cell substructure, biologists have turned to physical techniques for preparing biological specimens. Such techniques do not involve chemical fixatives. Two such physical techniques—freeze drying and freeze etching—are briefly discussed. This section is included only to introduce the techniques involved. References are given for more detailed accounts of these specialized procedures.

Freeze Drying

The freeze drying process was first shown to be useful for electron microscopy by Sjöstrand and Baker (1958). The process involves quick-freezing small units of tissue, vacuum distillation of all moisture from the tissue, and direct embedding of the tissue (via freeze substitution) with the appropriate plastic. Freeze drying is commonly used in histochemical studies and is discussed in detail by Sjöstrand (1967).

The tissue is rapidly frozen (to about $-72°$ C) in a dry ice bath and it is then placed in a vacuum chamber where the ice is allowed to sublimate. In some cases the tissue is first soaked in glycerol to prevent crystallization of water in the cells of the tissue. The tissue is then placed in a 1:1 solution of an acetone and plastic mixture or some other solvent. Embedding of the tissue then follows except that the tissue and plastic mixtures are kept at low temperatures until polymerization. Since both chemical fixation and normal dehydration through a solvent series can be avoided, leaching out of cell constituents will be reduced. The entire operation, particularly the freezing and drying (sublimation) of the fresh material and the plastic substitution in the tissue, should be completed as rapidly as possible.

MECHANICAL FIXATION 2.4

Freeze Etching

Freeze-etch replication for electron microscopy was first described by Moor (1964). It is most useful for the study of membrane structures. The technique involves the production of carbon-coated and platinum-shadowed replicas of frozen, fractured, and sublimated (etched) surfaces. A complete description of this process is given by Moor (1964, 1969), Bullivant and Ames (1966), and Stachelin (1968).

The specimens may be blocks of tissue, pellets of cells, or whole fragments of plants, and are infiltrated with glycerol or ethylene glycol (20%). The specimen may first be fixed in buffered glutaraldehyde. To avoid chemical fixation, the material can be frozen directly in 20% ethylene glycol in a veronal-acetate buffer. The addition of ethylene glycol is important for preventing the formation of ice crystals during the etching phase of the process. This cryoprotective substance also permits cleaving along protein surfaces, which aids in examination of cytomembrane systems.

Small fragments of the infiltrated specimen are then frozen in liquid nitrogen ($-196°$ C) and placed in a supercooled freeze-etch device. The entire unit is then transferred to the vacuum system. Prior to freezing the specimen, the carbon and platinum electrodes must be aligned to allow proper coating and shadowing of the fractured and etched surfaces of the specimen. The available commercial instruments are usually accompanied by instruction booklets. After attaining a vacuum of 10^{-4} torr (1 mm of Hg equals 1 torr of pressure) or better, the specimen is fractured, usually by a spring-loaded knife and the fractured surface is exposed to the vacuum. Usually the specimen is then warmed up through a built-in heat sink and, when the specimen reaches temperatures of $105°$ C, etching (sublimation of ice in a vacuum) will occur. Following a specified time of exposure (about 15 min in a vacuum of 10^{-5} torr will achieve a deep etch of the surface), the surface of the specimen is shadowed and coated with platinum and carbon. The replica is then removed from the specimen by dissolving the material in an organic solvent (bleach) and the floating replica is cleaned by transfer through solvents and examined in the electron microscope.

The technique avoids all the standard preparative techniques for sectioning material and yet allows examination of cell interiors.

3 DEHYDRATION

Introduction 3.1

Most embedding media are not miscible with water. Therefore, water must be removed and replaced with an organic solvent which will dissolve the embedding mixture. This is done by placing the tissues in increasing concentrations of alcohol or acetone. It is important that the process of dehydration be *thorough;* that is, all the water should be completely removed from the tissues so that polymerization damage does not occur.

Actually, the steps involved in dehydration are less of a problem to electron microscopists than other microscopists since the former work with very small specimens (of the order of 1 mm^3). Samples can be conveniently dehydrated by pipetting or pouring off most of the solution from a vial (5 to 10 cc) and adding new solution. The tissue should never be exposed to air during dehydration. Always leave a small residue of the previous solution covering the material. The more traditional dehydrating solutions are a graded series of ethanol or acetone from 10% to absolute (Fig. 3-1), in 10% or 30% steps for about 15 to 30 min at each step at 4° C (in an ice bath). However, as can be seen in the dehydration schedules, a wide variety of schedules and dehydrants are employed.

The temperature at which the dehydration process takes place is important. Most present-day schedules call for low or even freezing temperatures. Low temperature dehydration is thought to prevent lipid extraction.

Tissues fixed with osmium tetroxide must be thoroughly washed before dehydration in an alcohol series. Any osmium present in the tissue, which has not reacted with cell constituents, will react with the alcohol, forming black colloidal particles of reduced osmium. Rinsing should consist of at least three 15 min rinses followed by a 1–2 hr rinse in the appropriate buffer in the cold. Acetone apparently will not react with osmium in the same fashion as alcohol.

3 DEHYDRATION

Figure 3-1 A demonstration of a complete dehydration series of 10% steps in ethanol, three transfers in propylene oxide-ethanol (1:3, 1:1, and 3:1), and then 100% propylene oxide.

There is considerable variation in the length of dehydration. Some investigators emphasize the necessity for rapid dehydration, to prevent leaching out of material which can occur if dehydration is too slow. However, other investigators feel that dehydration can be either slow or rapid, but they emphasize that the graded concentrations of solvent be small. Still other investigators have found that large steps of increasing alcohol concentrations, taken at a leisurely pace, are successful.

All investigators agree that dehydration must be complete. If moisture remains in the tissue, polymerization will be inhibited and the material will not section well. To prevent this, several changes of absolute solvent should be carried out before infiltration with the plastic mixture. The solvent itself must also be water-free. This can be accomplished by the addition of some hygroscopic, insoluble substance like sodium sulfate (Na_2SO_4) to the absolute alcohol or acetone.

SCHEDULES 3.2

Aspiration may be required during dehydration to remove trapped air bubbles, especially in plant cell walls. This should be done at intermediate stages in the dehydration process, that is in the 30 to 50% range. A good rule of thumb for determining the need for aspiration is to note whether the tissue floats in the vials during dehydration. If a faucet suction device is the source of vacuum, the process of aspiration should be carefully watched to insure that no back-up of water into the desiccator takes place.

Finally, it is sometimes recommended to use transitional solvents after the use of alcohol in the dehydration process. This is an especially common practice in epoxy embedding, where the practice arose because of nonuniform impregnation by the plastic when ethanol was used. Luft (1961) recommends the use of propylene oxide as a transitional solvent between the alcohol and the plastic or resin mixture. He demonstrated that propylene oxide is reactive with the plastic so that its molecules become incorporated into the structure of the polymerized plastic.

Not all investigators agree with this approach and many prefer to avoid the use of propylene oxide altogether. Some disadvantages of propylene oxide include its highly volatile nature, somewhat toxic character, and its possible influence on the qualities of the plastic. This solvent should be used only in a fume hood. Acetone has been recommended as a single solvent in dehydration, substituting for both ethanol and propylene oxide (see schedule (3)).

The tissue can be stored in the refrigerator in 95% ethanol or 90% acetone for days, although it is usually recommended that infiltration and embedding in plastic should follow dehydration. The processes of infiltration and embedding will be the subject of the next chapter.

Dehydration Schedules 3.2

In the preparation of the various concentrations of the alcohols, the use of more costly absolute alcohol as the basic solution for dilution should be avoided. For example, 95% ethanol is used rather than absolute. Dilution of the 95% solution is a simple matter. The amount, measured in milliliters of 95% ethanol equal to the desired concentration, is poured out, and to this amount sufficient distilled water is added to make up 95 ml of solution. For example, a 30% ethanol solution is made by taking 30 ml of the 95% ethanol solution and adding 65 ml of distilled water to make the final amount (95 ml) of solution. The following dehydration schedules are described in detail as to their preparation and use.

Ethanol-Propylene Oxide—Slow Schedule

This schedule is a modification of the one used by Luft (1961). Dehydration is performed in 10% steps of ethanol with 15 min soaking at each step interrupted by aspiration during the 30%, 40%, and 50% steps (Fig. 3-1), and all are done in an ice bath. After two 15 min changes in dry absolute alcohol, the tissue is then transferred into propylene oxide for 30 min each in solutions of the following proportions: 1:3, 1:1, and 3:1 propylene oxide to ethanol. Finally, the tissue is held for two 30 min changes in 100% propylene oxide.

This rather slow dehydration schedule appears to be necessary for some types of plant material that will collapse with too rapid a dehydration schedule. Aspiration is recommended at the intermediate steps to insure that air bubbles trapped between plant cell walls are removed.

Ethanol-Propylene Oxide—Rapid Schedule

Many investigators wish to avoid prolonged exposure of the tissue to solvents which may result in swelling of the tissue. They recommend starting dehydration at the 30% level (ethanol) and then going through 50%, 70%, 90%, and 100% ethanol, and then directly into 100% propylene oxide. Each step should last about 30 min and all steps are carried out in an ice bath.

Acetone Dehydration

Although Luft (1961) recommended the use of propylene oxide as a transitional solvent for infiltration in epoxy resins, some investigators argue that propylene oxide does cause irregular polymerization in the block. Therefore a schedule using only acetone is presented here. Transfer the fixed and rinsed tissue into 30% acetone for 30 min to 1 hr at 0° C. Following this, transfer the tissue through 50%, 70%, 90%, and two changes of 100% dried acetone (add Na_2SO_4 to acetone). Then bring the tissue to room temperature in the infiltration mixture of 1:1 acetone to plastic mixture. When working with cold solutions, special care must be taken to avoid condensation of moisture inside the vials.

Infinite Dilution

Add, drop by drop, absolute alcohol or acetone until the volume is twice that of the initial amount (50%). Then pour off one half of the mixture and repeat the process so that a concentration of 75% acetone or ethanol is obtained. Finally pour off all but a thin covering layer of solution and add dry absolute ethanol or acetone. This procedure permits the dehydration of delicate tissue without handling the tissue. This material can be transferred into propylene oxide if desired.

Other Series

Use of butanol-xylene or ethanol-xylene have been recommended in the past. However, these solvents are used to a much greater extent in light microscopic preparations today than in electron microscopic procedures.

4 EMBEDDING

Introduction 4.1

Ultrathin sectioning could not become a simplified, reproducible technique until suitable media were introduced in which the material could be properly embedded. Waxes, which are common embedments in light microscopy, are neither durable enough to permit ultrathin sectioning (von Ardenne (1939)) nor could they withstand the intense heat of an electron beam. Pease and Baker (1948) adapted parlodion from light microscope use as the first plastic embedment. It was used as a coat around the wax embedded tissue and consequently the section was still sensitive to the electron beam.

Newman *et al.* (1949) introduced n-butyl methacrylate which is easily sectioned and is soluble in many dehydrants. Unfortunately, this plastic is also beam sensitive, and furthermore methacrylates cause some polymerization damage and the plastic will shrink up to 20% during hardening, thus distorting cell structure. Methacrylate was the most commonly used plastic until around 1956. At that time Maaløe and Birch-Anderson (1956) introduced the first epoxy resins for embedments in electron microscopy. These plastics are very beam stable (thermostable) and do not cause polymerization damage during hardening. The epoxy resins have certain faults. They are quite viscous, making them difficult to handle and are also more electron dense which unfortunately reduces specimen contrast in image formation.

Presently, a variety of substances are used as embedments including epoxy resins (such as araldite, epon, Maraglas, and Durcapan), a few methacrylates (such as n-butyl and n-methyl methacrylate), and a polystyrene plastic (such as Vestopal). Recently Spurr (1969) described a new epon based plastic mixture which, in addition to the desirable qualities of epoxy resins already listed, has the added benefits of being low in viscosity and quite transparent to the electron beam.

Regardless of the plastic used, it should meet the following criteria:

1. The fine structure of the embedded tissue should be free from polymerization damage by the epoxy resin.

4 EMBEDDING

2. The plastic should be stable when exposed to the electron beam.
3. The degree of electron scattering caused by the embedment should be minimal so that contrast of the specimen in the microscope is not reduced.
4. The plastic should cut well. Therefore both the block hardness and hydrophobic characteristics must be consistent.
5. The plastic mixtures should be of low enough viscosity to permit accurate measurement of the components of the mixture and to allow proper infiltration.

4.2 Handling of the Plastics

Epoxy resins used as embedding media require more specialized handling than waxes and other embedments. Since all of the epoxy resins or their accelerators (catalysts) may cause dermatitis or even be carcinogenic, it is recommended that plastic gloves be used and that work be carried on in fume hoods. All spilled resins should be wiped up with a solvent-soaked towel and care should be taken to avoid touching any objects (switches, knobs, etc.) when working with these mixtures. A covered waste container should be adjacent to the fume hood for discarding contaminated materials.

With the exception of a new, low viscosity epon described by Spurr (1969), the epoxy mixtures are typically viscous. These require thorough mixing to insure that proper polymerization will take place. The general mixing procedure followed should be a pumping action with a clean glass rod carried on for up to 15 min to insure adequate mixing.

The two common methods of measurement have been volumetric and weighing. Weighing is preferred because of the difficulty of accurately determining volumes. If volumetric measurement is used, spilling of resins on the sides of the plasticware should be avoided. Accurate delivery of the accelerator (resin catalyst) is most important and a pipette is recommended.

Disposable plasticware is recommended for measuring and mixing of the resin mixture because of the great difficulty in cleaning glassware. After use, the plastic beakers, graduated cylinders, and glass rods should be drained and placed in an oven to permit polymerization of the remaining plastic. Some of the plasticware usually can be salvaged simply by tapping the containers on a table top which often loosens the polymerized plastic. Adding a small amount of complete resin mixture to each resin-contaminated container will insure polymerization of all remaining plastic since unmixed resins will not polymerize. Be cautious with plastic gloves and beakers which may dissolve in the plastic resin and cause poor polymerization.

If glassware is used, a cleaning solution should be available for removal of the plastic resins. An excellent soaking solution can be made by using Decontam (20 cc/quart water, Kern Chemical Corp., Appendix I). Glassware will be freed from contamination with polymerized or unpolymerized resins in about 6 to 24 hr in this solution (procedure used by Myron C. Ledbetter, Brookhaven National Laboratories).

Infiltration 4.3

The plastic or resin mixture is next infiltrated into the specimen, which at this point is in an absolute solvent or a transitional solvent. Infiltration is a most critical process since poorly polymerized tissue results if the resin mixture is not well infiltrated. Special care should be taken to avoid water contamination of the resins. A good procedure is to place the tissue in a 1:1 ratio of dry solvent to full resin mixture including the catalyst (accelerator), and then to follow this with at least two changes in a 100% plastic mixture at room temperature. Each of these steps should last from 2 hr to overnight. Short exposures to higher temperatures can be used to lower the viscosity. Partial polymerization of the plastic mixture may occur and therefore scrutiny of the heated specimen-plastic mixture is required.

At each step in the infiltration process, the tissue and resin mixture may be placed in a desiccator and a weak vacuum applied. This procedure removes any trapped air bubbles formed during dehydration. Such treatment is especially useful in handling plant tissue because the cell wall is difficult to embed. If the tissue is floating at any time during dehydration or infiltration, aspiration is recommended. If available, a vacuum oven is most useful for this purpose. Although shakers and rotators aid in infiltration, the movement must be gentle or damage to the tissue may result.

Capsule Embedding 4.4

Both gelatin capsules, obtained from local supply houses (size 00) or polyethylene capsules (available from supply houses, Appendix I) are useful. These capsules can also be used for the infil-

4 EMBEDDING

Figure 4-1 Materials used in embedding are shown: gelatin and Beem capsules, plastic syringe, scissors, tweezers, toothpicks, and dissecting needle.

Figure 4-2 The process of injecting the plastic mixture into a capsule using a syringe (10 cc) minus needle. First, a small drop of plastic is placed in the capsule, then the specimen is transferred to the capsule, and finally the capsule is filled as shown.

CAPSULE EMBEDDING 4.4

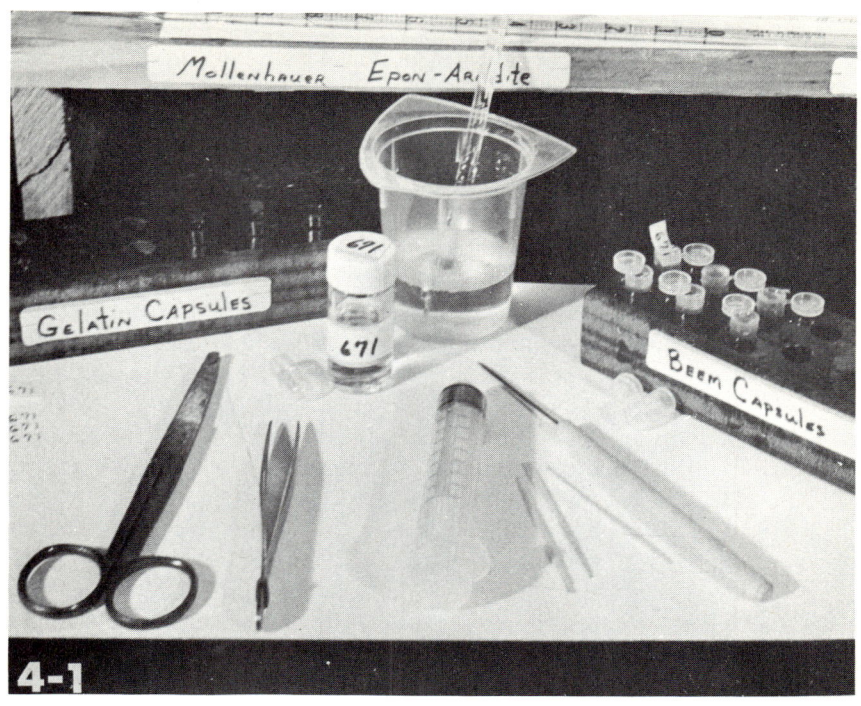

4-1

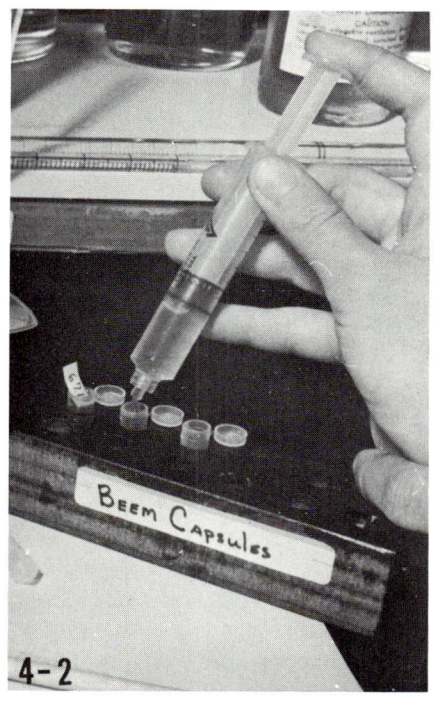

4-2

4   EMBEDDING

tration steps rather than using the vials. If gelatin capsules are used, insure that they are oven dry before use, otherwise poor polymerization of the plastic will result.

The tissue is simply transferred from one capsule to another by a toothpick (Fig. 4-1). A small drop of plastic solution is placed at the base of the capsule, which is held in a capsule holder. The tissue is transferred into this capsule after a brief draining on filter paper to remove excess solution from the previous soaking. The capsule is next filled to the top with additional resin mixture. Plastic mixtures can be injected into the capsules by using a disposable 5 or 10 ml syringe without a needle (Fig. 4-2). If necessary, to insure that the tissue will not float to the surface during the final embedding, a pin or fragment of a toothpick can be pressed against the tissue, forcing it to the capsule base. A label may then be placed in the capsule for identification, and the holder with capsules placed in the oven for polymerization. The gelatin capsules are removed by soaking in hot water or chipped off with a razor blade (Fig. 4-4). Beem capsules can be stripped off by making two longitudinal cuts in the capsule (Fig. 4-4). Capsule holders can be made by drilling holes in a 2- by 4-inch board or similar material.

Tissue flotation just before polymerization is common during the embedding process. This occurs when the resin mixture reaches its lowest viscosity, and a modification of the capsule technique is described here (Fig. 4-3, AIBS handout, 1967).

The equipment needed for this procedure is:

1. a glass plate suitable to fit in the polymerization oven;
2. tools for handling the tissue (forceps, wooden applicators, etc.);
3. gelatin capsules with labels;
4. plastic syringe and the embedding resin mixture; and
5. Scotch Double Tape #665.

Strips of tape are placed across the glass plate, and the infiltrated tissue, which is ready for embedding, is positioned on the tape about 2 cm apart. The tissue is oriented with the area to be sectioned facing down. The gelatin capsules are then filled to the brim with the resin mixture and are then quickly overturned on the tissue and centered. To check that the orientation of the tissues remains the same, hold the glass plate up to an overhead light and examine from underneath. Take care to avoid tipping the individual capsules. If the orientation was lost, the capsule may be removed and the tissue reoriented. The glass plate is then placed in the oven and the resin is polymerized. The tape strips can be peeled off the glass plate, and the capsules will come off the tape easily.

Flat Embedding 4.5

This has been suggested as an alternative to capsule embedding because of its better preservation of the tissue structure and its more consistent cutting properties. Furthermore, flat embedding is a faster process since a number of specimens may be embedded in the same dish, and then only one label is needed. If orientation in capsules is a problem, because of size or shape of the tissue, flat embedding may be used (Fig. 6-1).

Embedding dishes can easily be made by using the tops of large mouth polyethylene bottles or simply by using the basal portion of a 50 ml disposable polyethylene beaker. In each case, the resin mixture is poured into the container to a level of about 5 to 8 mm, and the tissue oriented around the edges of the beaker. Better orientation can be achieved if the resin mixture is warmed in an 80° C oven for 5 to 10 min prior to pouring. A small label is then placed in the center of the beaker. The beakers containing tissue and resin mixture are next placed in the oven and after about 15 min are checked to see that the specimens are still oriented as desired. The advantage of plastic disposable beakers is that removal of the plastic block requires only a tapping of the lip of the beaker to loosen the block (Fig. 6-1). The beaker can then be reused.

C. W. French (Appendix I) manufactures flexible silicone rubber molds for flat embedding. Once the epoxy mixture has polymerized, the blocks are easily released from the mold. These blocks have the correct size for all microtome chucks and a single mold holds twenty separate specimens.

Particulate Specimen Embedding (Pre-embedding) 4.6

A pre-embedding method is used on free cells and micro-organisms in order to obtain cubes of material which can be handled like pieces of tissue in dehydration, infiltration, and embedding. The pre-embedding technique forms pellets of cells which might otherwise tend to disintegrate. Fixation, dehydration, and embedding in plastic can then be easily performed.

4 EMBEDDING

Figure 4-3 A demonstration of the inverted capsule technique to permit buoyant specimens to rest at the capsule tip.

Figure 4-4 A demonstration of the Spurr epon plastic mixture which uses gravimetric techniques (scale in upper portion of photograph). In the foreground, the soaking of gelatin capsules to remove the gelatin and the stripping off of the Beem (polyethylene) capsules are shown.

PRE-EMBEDDING 4.6

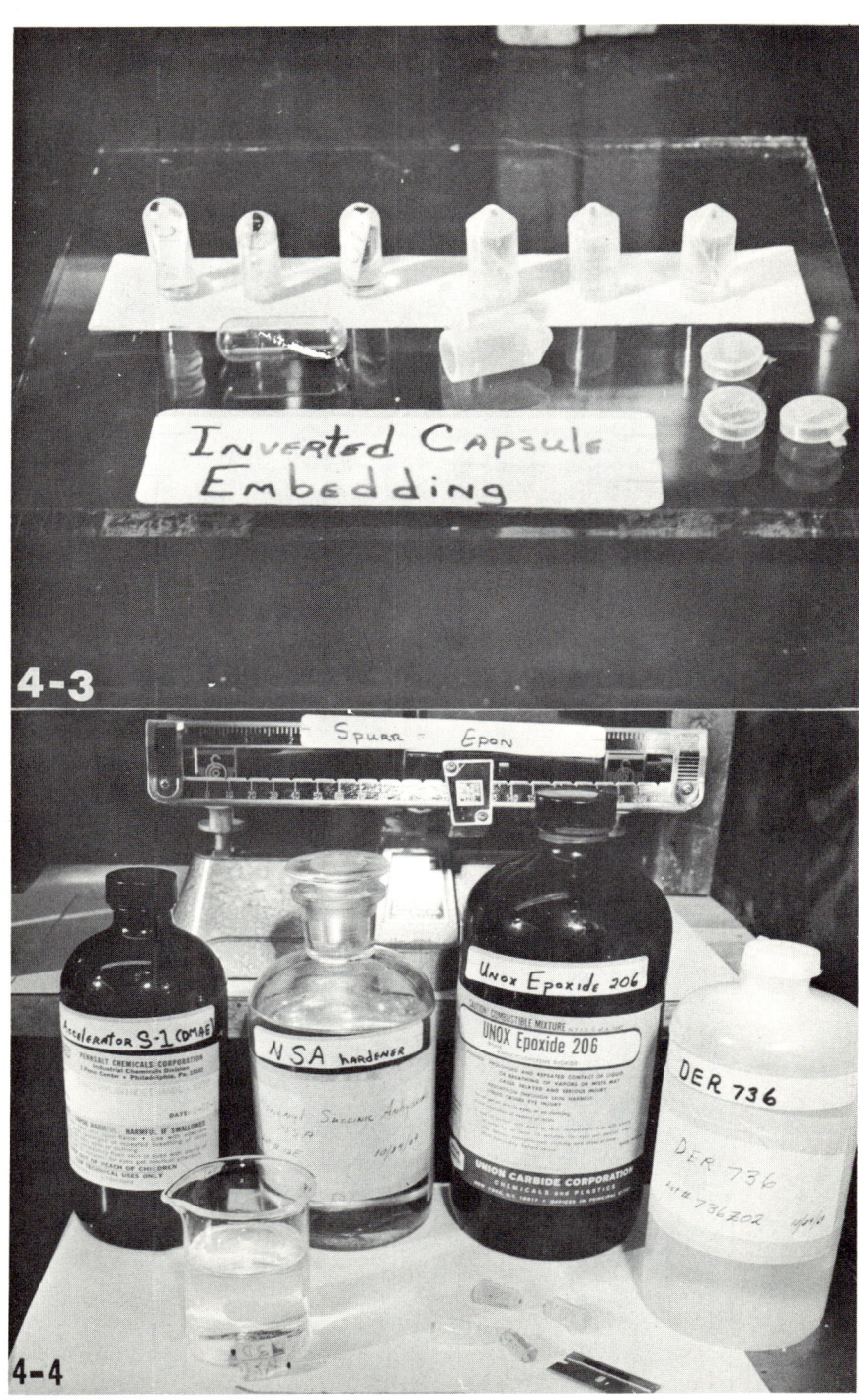

4 EMBEDDING

Any fixative can be used to fix the free cells. The cells are usually fixed by suspension in the fixative or by addition of the fixative to the cell suspension. The time of fixation varies according to the specimen, but usually it is short and lasts for 10 to 15 min at room temperature. It may be desirable to rinse the specimen free of the fixative before actual pre-embedding. This can be done by centrifugation and resuspension in a buffer solution and then followed by a second centrifugation. There are two common materials used in pre-embedding—agar and bovine serum. A new technique using a millipore filter is described by Thomas and Maher (1969).

Agar Pre-embedding

This is usually carried out in a 4% agar concentration. Place 0.4 cc of dry powdered agar in a graduated 10 ml cylinder and add water to the 5 ml mark. Heat the mixture in a water bath at 100° C and completely dissolve the agar granules by stirring for 10–20 min. An 8% agar solution results. Formalin, saline, sucrose, or buffer solutions may be used instead of distilled water for making the agar solution.

Pre-embedding of pellets of free cells or micro-organisms which have been fixed and centrifuged is carried out as follows. Suspensions of the specimen concentrated in a centrifuge tube are mixed with an equal volume of the 8% agar solution at a temperature of 46°–47° C forming a 4% agar solution. After gelation of the pellet-agar block in the centrifuge tube, the block can be removed by addition of a few drops of 70% ethanol. The ethanol causes a slight shrinkage of the block which can then be removed with a needle.

Material embedded in agar blocks can then be postfixed in osmium if desired or in formalin if the agar block needs toughening. The cubes cut from the agar block can then be dehydrated as with other specimens and embedded directly.

Bovine Albumen Pre-embedding

Pre-embedding with bovine albumen is carried out in a manner similar to the process described above for agar embedding. A 1% solution is made and mixed with the cells and a drop or two of 5% glutaraldehyde is then added to coagulate the protein. This suspension is centrifuged and the albumen block removed. This technique is not recommended if tissue is to be postfixed with osmium since a reaction with the albumen will take place.

5 PLASTICS

Introduction 5.1

As noted in the previous chapter, there are two major types of plastics used as embedments—methacrylates and epoxies. A third material—polystyrene—is also used but on a lesser scale. The sources and proper chemical names of the plastics are given in Appendix I. There are a number of water soluble methacrylates and epoxy resins which do not require dehydrants in embedding. These will not be discussed here as the results in the author's laboratory have been quite variable and in some cases not satisfactory. The reader is referred to the information available from Polysciences and Ladd, Inc. (see Appendix I).

Methacrylate 5.2

The most satisfactory methacrylate embedding mixture for electron microscopy is a mixture of n-methyl methacrylate and n-butyl methacrylate. Each of the liquid monomers should be cleared of the inhibitor. The inhibitor is usually 1% hydroquinone which prevents autopolymerization of the liquids during storage. Pease (1964, p. 98), however, does not feel that removal of the inhibitor is necessary. The two monomers are then mixed together in the desired proportion, and a catalyst is added to speed up polymerization. Polymerization may be accomplished in an embedding oven, under atomic radiation, or under ultraviolet light.

When used correctly, methacrylate plastics give excellent results. They have good cutting properties, are very electron transparent, and permit stain penetration in the embedded tissue. However, there are a number of drawbacks with methacrylates. First, there is a tendency toward compression thus distorting the specimen. A more difficult problem is disruption of cell structure with the use of methacrylates due to polymerization damage from shrinkage of the plastic during hardening (up to 20% shrinkage can occur). Polymerization damage can be reduced by thorough

5 PLASTICS

dehydration, postfixation in formalin, addition of 1% uranyl acetate to the lower alcohols during dehydration, and prepolymerization of the plastic mixture prior to embedding. Another unfortunate characteristic of methacrylates is their sensitivity to the electron beam. Methacrylates will evaporate under the intense electron beam and the vapors will condense on electron microscope apertures causing contamination. This evaporation or etching of the plastic actually improves contrast in the specimen, but will also result in beam damage to the specimen. This problem can be reduced by coating the sections with evaporated carbon.

Plastic Mixture and Embedding

To insure good polymerization it is recommended that the inhibitor be removed from the two monomers (n-butyl methacrylate, n-methyl methacrylate). The inhibitor can be removed by mixing 1 part of 5% NaOH and 2 parts methacrylate (either monomers can be used) in a separatory funnel. Shake this solution and let stand for 15 min, then drain off the purple-colored NaOH and hydroquinone layer on the bottom of the funnel. Repeat this process (about 3 times) until the NaOH solution is clear. Wash with distilled water to remove all NaOH (the drain water should no longer feel "slippery"). Place the monomers in glass stoppered bottles, add $CaCl_2$, let stand overnight, filter and repeat filtration after one more night (use #42 filter paper). The two monomers will keep in the refrigerator for about a year. Always bring the plastic monomers to room temperature before using to avoid water condensation. Avoid skin contact with the monomers.

The hardness of the final blocks depends mainly on the ratio of the two methacrylates in the mixture. For harder blocks, the proportion of n-methyl methacrylate is increased, while for softer blocks, the proportion of n-butyl methacrylate is increased. A 1:2 ratio of n-methyl-n-butyl mixture is good. This mixture usually will last about a year (with the inhibitor removed) in the refrigerator. Dalton (unpublished notes) recommends 0.25% benzoyl peroxide, and Sjöstrand (1967) recommends 0.1–1% catalyst. Dissolve the catalyst in a small amount of the mixture, then add the remainder of the plastic. This mixture will keep in the refrigerator for 2 months before polymerization. Benzoyl peroxide will keep at room temperature for a year.

Infiltration: After a careful dehydration in an ethanol series, the material can be infiltrated directly into the monomer mixture of methacrylates. Take 1 part absolute alcohol and 1 part of the plastic mixture and allow 2 changes of 15 min each for the specimen. Then proceed with 100% plastic mixture, adding catalyst as required, in a series of steps:

> Mixture, 2 changes, 15 min each, overnight in refrigerator if desired.
> Mixture plus catalyst, 15 min.
> Mixture plus catalyst, overnight in refrigerator.
> Mixture plus catalyst, in capsule, in 60° C oven, overnight.

Some investigators partially pre-polymerize the monomer mixture with the catalyst used for the final embedding in order to help reduce polymerization damage. One method is to heat the capsules containing monomer mixture with catalyst in the oven for 15 to 50 min, until it reaches the consistency of honey. Specimens are then placed in the capsules, and polymerization is completed as usual. In an alternate method (less likely to produce trapped air bubbles), the monomer mixture with catalyst is heated in a sealed tube, and the tube is inverted every 15 min until the plastic reaches the consistency of honey. A rubber bulb can be attached to a broken end of this tube, and the other tip is then broken. A drop of the prepolymerized plastic is placed in each capsule. The specimens are placed into the drop at the end of each capsule, and the capsules are filled with plastic. Polymerization is completed as usual.

Epoxy Resins 5.3

A number of different epoxy resins are presently being used as embedments in electron microscopy. The reasons for the popularity of these plastics is their thermostability with little or no shrinkage upon polymerization (not more than 1.5%) and their good cutting qualities. However, the resin mixtures are viscous, toxic (even carcinogenic), and difficult to handle.

A number of commercial types of epoxy resins exist; their components, however, are similar. They are an epoxy resin (one or more required per mixture), hardeners, accelerators, and plasticizers. A discussion of the chemistry of each of these can be found in Appendix II. Several recommended epoxy formulae and their specific uses are listed below.

5 PLASTICS

Epoxy Resin Formulae

(1) Glauert and Glauert (1958), English araldite In general, araldite epoxy resin mixtures are harder than epon resins and have been very popular for cutting hard, gritty specimens (pollen grains, calcium carbonate coated algal cells, and cells with crystals (crystoliths)). Another advantage of araldite plastics is that sections of material embedded in this plastic can be "spread" out with toluene vapors.

Components: The English araldite consists of four components, which can be kept at room temperature.

araldite resin M	
(can also use araldite 6005)	10 ml
hardener 964B	
(can also use DDSA, see below)	10 ml
dibutyl-phthalate	1.0 ml
accelerator 964C	
(can also use tridimethaminomethyl phenol)	0.5 ml

Equal parts of araldite and hardener are first mixed in a beaker. Dibutyl-phthalate is added and then the accelerator is mixed in. Since the final hardness is determined by the amount of dibutyl-phthalate added, the amount should be varied to obtain the desired hardness of plastic.

Infiltration and Embedding: Infiltration is accomplished by first mixing a 1:1 solution of plastic mixture (without accelerator) and acetone or propylene oxide and then five to six changes at 1–2 hr per change of 100% plastic mixture without accelerator at room temperature. Final embedding is done using a complete plastic mixture (with accelerator). This mixture and tissue are left at room temperature for 2–3 hr, and then placed in fresh complete plastic mixture in capsules in the embedding oven (48° C) for about 30 hr.

(2) Luft (1961), American araldite

Components: The American araldite mixture consists of the following three components to which the fourth can be added if desired.

araldite 502 (Araldite M or 6005)	27 ml
hardener DDSA	23 ml
accelerator DMP-30	(1.5–2%)
dibutylphthalate	up to 2% is desired

The above formula is slightly modified from that given by Luft (1961). The major difference from the English araldite mixtures

EPOXY RESINS 5.3

of Glauert and Glauert (1958) is the ratio of hardener to epoxy resin. Luft has emphasized that nonreactive solvents should not be left in the tissue during the embedding or only partial polymerization will result.

Infiltration and Embedding: Infiltration is carried out in a 1:1 mixture of complete plastic: propylene oxide overnight, followed by two 30 min changes in pure, complete plastic mixture at 48° C and finally embedded in a third change at 60° C overnight. To prevent polymerization during the two 30 min changes at 48° C, remove the plastic from the oven and check for viscosity a few times during each step.

(3) Cargille (1962), American araldite The araldite mixture of Cargille (kits with all components can be purchased from Cargille, see Appendix I) is based on the English formula (Glauert and Glauert (1958)). The ratio is different from that proposed by Luft (1961) and a slightly harder plastic will result. When compared to the formula of Glauert and Glauert (1958), the ratio is almost identical in the final mixture (solution III).

Components: Three solutions are required for infiltration and embedding.

	I	II	III
epoxy resin A (Araldite M, 502 or 6005)	5 ml	5 ml	10 ml
hardener DDSA (dodecyl succinic anhydride)	5 ml	5 ml	10 ml
dibutyl-phthalate (if desired)	...	1–3 ml	1–3 ml
accelerator B (n-benzyldimethyl-amine)	0.5 ml	...	0.5 ml

Infiltration and Embedding: After dehydration, the tissue is soaked in a mixture of solution I and absolute alcohol (propylene oxide can be used here as a transitional solvent, and is recommended in this text), for 3 hr at 48° C, or overnight at room temperature. Occasional gentle stirring with a toothpick prevents tissue from sticking to the bottom. After draining on filter paper, the tissue is transferred into solution II and can be left overnight at 48° C. This step invariably results in polymerization of the tissue unless there has been thorough draining of solution I. It seems best not to place the tissue and solution II in an oven, but rather leave it out at room temperature overnight. Occasional stirrings are recommended during this step. Upon completion of solution II, the tissue is transferred after draining into solution III. Note that the amount of accelerator in solution III is exactly $\frac{1}{2}$ that of solution I. This stage of infiltration (solution III) should last up to 20 hr at room temperature with the solution being preheated to 48° C. Finally, embedding follows after draining the tissue and placing it in fresh solution III at 48° C overnight. The capsules or beakers should be examined and the specimens reoriented after about 15 min.

(4) Luft (1961), Epon Epon resins have become very popular because of the softer plastic and ease of cutting when compared to araldite. These plastics are more electron transparent, offering a higher contrast to specimens embedded in them. However, with the exception of the newest epon by Spurr, most epon resins are more viscous and difficult to handle than the araldite resins. Sections of epon embedded material do not expand as well with toluene of chloroform vapors.

Components: The individual components may be kept in stoppered bottles at room temperature, but the two solutions, A and B, should be kept in the refrigerator until needed (up to 6 months). Allow solutions to come to room temperature before using to avoid water condensation.

	A	B
epon resin 812	62 ml	100 ml
hardener DDSA (dodecyl succinic anhydride)	100 ml	...
hardener NMA (nadic methyl anhydride)	...	89 ml
accelerator DMP-30, (2,4,6, dimethylaminomethyl phenol)		1.5–3%

Immediately before use, the accelerator is added to the resin mixture (1.5% up to 3% as the accelerator ages) and stirred thoroughly. Note that two different hardeners are used and that no plasticizer is required because of the softer plastic.

Infiltration and Embedding: The specimen is dehydrated according to schedule and brought into 100% propylene oxide. It is then placed in a 1:1 mixture of plastic resin and propylene oxide for 1 hr to overnight at room temperature with occasional swirling to prevent the tissue from sticking to the bottom. After draining on filter paper, the specimen is transferred into fresh plastic mixture for about 30 min at room temperature. This is repeated at least once and can be allowed to occur under a weak vacuum to remove trapped gas bubbles. Embedding is accomplished by draining the specimen on filter paper and placing it in fresh plastic mixture (in capsules or beakers) in a 60° C oven for overnight polymerization.

Note: Blocks of various hardness can be obtained by using different proportions of solutions A and B. Pure mixture A will produce a soft block, while pure mixture B will produce a hard block. Some laboratories (including the author's) use only solution B with Epon 812 and NMA in a 1:1 formula which produces a very hard block. The recommended mixture (solutions A and B) is as follows:

solution A	7	ml
solution B	3	ml
DMP-30	0.15 ml	

(5) *Spurr* (1969), *Epon* This epoxy resin mixture may possibly terminate the long progression of attempts to obtain an epoxy of high electron transparency, low viscosity, and thermostability. Such characteristics permit high specimen contrast, ease of penetration of the plastic during infiltration, and stability of the sections in the electron beam.

The plastics are available as separates (see Fig. 4-4) or as a kit (see BES or Polysciences, Appendix I). The plastic mixture is prepared by the gravimetric rather than the volumetric method insuring higher accuracy in measurement. Each ingredient is added in the above order to an Erlenmeyer flask placed on a balance. Mixing is accomplished by swirling the flask after each addition. The total solution can be stored in the refrigerator for up to 7 days.

Components: The components of the plastic mixture are:

UNOX Epoxide 206 or ERL 4206 (vinyl cyclohexene dioxide)	10.0 g
hardener NSA (nonenyl succinic anhydride)	26.0 g
plasticizer DER 736 (diglycidyl ether of propylene glycol)	6.0 g
accelerator S-1 (DMAE: dimethylaminoethanol)	0.4 g

Infiltration and Embedding: The material is dehydrated using any standard dehydrant desired (ethanol, propylene oxide, butyl alcohol, or acetone). A 1:1 solution of absolute alcohol (or other dehydrant) and complete plastic mixture is made up and the tissue placed in it. Half of this mixture is poured out after

5 PLASTICS

30 min at room temperature and an equal amount of pure plastic mixture added. After 30 min pour off and replace with pure plastic mixture. The specimens are now in pure plastic and this step may be held until infiltration is complete (2–3 hr for small specimens, overnight for larger ones). The tissue may be transferred into fresh mixture and left at room temperature overnight and under vacuum if infiltration is a problem. Embedding is accomplished by transferring the material to capsules and a fresh plastic mixture in a 70° C oven overnight (about 8 hr).

Note: Block hardness will depend upon the amount of DER 736 used in the plastic mixture and this can be varied. The lower the amount, the harder the plastic. Use oven dry capsules and capsule holder, otherwise the bases of the castings become quite brittle (because of the presence of moisture in the gelatin capsules).

(6) *Mollenhauer (1963), Epon-araldite mixture* Although these mixtures were developed by Mollenhauer for embedding plant tissue, especially the plant cell wall, they have found great success in embedding animal and human tissues. The tissue preservation is comparable to that obtained when using epon, but the blocks are easier to section than epon or araldite.

Components: The components are varied according to which mixture is selected. Mixture #1 yields the softest plastic, being the easiest to section with glass knives. Mixture #3 yields the hardest plastic and is excellent for embedding soft, friable or brittle tissues (e.g., pollen grains, agar embedded cell suspensions). Mixture #2 is recommended for most routine plant and animal tissues and is the common mixture used in the author's laboratory.

	1	2	3
epon 812	25 ml	62 ml	...
araldite M (araldite 502, 6005, epoxy resin A)	15 ml	...	...
araldite 506	...	81 ml	50 ml
hardener DDSA	55 ml	100 ml	...
hardener cardolite NC-513	...	...	25 ml
plasticizer (dibutyl-phthalate)	2–4 ml	4–7 ml	1–2 ml
accelerator (DMP-30)	use 1.5% to 3%, depending on age		

The procedure for measurement is usually volumetric although almost as accurate results can be obtained by using gravimetric procedures. Simply substitute grams for milliliters so that the various solutions can be poured into a beaker on a top loading balance. The plasticizer and accelerator should be carefully pipetted into the beaker containing the already mixed plastic-hardener solution. Thorough mixing is necessary to insure proper polymerization. The amount of accelerator (DMP-30) can be varied according to age. Older accelerator is less reactive and consequently more should be used (up to 3%). Accelerator should not be kept for more than 6 months after opening the bottle. We have found the plasticizer (dibutyl-phthalate) is not necessary and can be omitted.

Infiltration and Embedding: Infiltration of thoroughly dehydrated tissue should be initiated in a 1:1 solution of propylene oxide and the complete plastic mixture (formula of choice). This initial step can be done at room temperature overnight with occasional stirring to prevent tissue sticking to the bottom of the vial. Two infiltration steps with pure plastic mixture should follow, each step of about 1 hr duration at room temperature. Embedding can then be accomplished by transfer of the tissue, after thorough draining on filter paper, into fresh plastic mixture in capsules or beakers. Polymerization is accomplished in an 80° C oven overnight. A longer time in the oven has no adverse effects. We have found that flat embedding appears to give better results in some cases than capsule embedding (ease of cutting and tissue preservation).

(7) *Freeman and Spurlock* (1962), *Maraglas mixture* This epoxy is available in kit form from Polysciences, Inc. (see Appendix I). The plastic blocks produced from this mixture have excellent cutting qualities and rather high electron transparency. However, this plastic is rather viscous and consequently it is similar to the epon plastics. The resin before cure is compatible with propylene oxide and acetone, but not with ethyl alcohol.

Components: Typical compositions are shown in the table below as reported in the information sheets published by Polysciences, Inc. The column marked "standardized media" is considered to be the best general purpose mixture for embedding of biological specimens and has been used in the author's laboratory.

Two modifiers are used to change the properties of the basic Maraglas epoxy resin. One is cardolite NC-513, which is a long

5 PLASTICS

Mixture:	A	B	C	D	E	F	stand-ardized media
Maraglas 665	60%	60%	65%	65%	70%	75%	68%
cardolite NC-513	40	30	30	20	20	15	20
dibuthy-phthalate	..	10	5	15	10	10	10
benzyldimethyl-amine (BDMA)	2	2	2	2	2	2	2

chain monomeric epoxide of amber color. It has a low viscosity and therefore acts as an internal non-extractable plasticizer. Dibutyl-phthalate is an external plasticizer used to give a hard but easily cut embedment suitable for sectioning.

Infiltration and Embedding: The specimen is prepared in the traditional manner for epoxy embedment by graded dehydration into 100% propylene oxide. After removal of the tissue from propylene oxide, it is graded in two changes of 1:1 propylene oxide-Maraglas mixture, 30 min each. The tissue is then impregnated with 100% Maraglas mixture for 30 min at 10° C and then allowed to come to room temperature. Finally the tissue can be embedded in predesiccated capsules with fresh mixture in a 60° C oven for 24 hr.

Note: Sometimes tissue impregnation is incomplete, as when soft tissue is surrounded by a hard matrix. In such cases, a modification of the original procedure is recommended in which a slower infiltration-embedding sequence with more plastic changes is followed.

(8) Ryter and Kellenberger (1958), Vestopal This polyester is an acrylic resin which has enjoyed wide use, especially in Europe. It is a softer plastic for cutting than the epoxies. The plastic is usually used in a partially polymerized form and consequently is rather viscous and difficult to handle. Also it is not miscible in some solvents (ethanol) although it is soluble in acetone or styrene.

Components:
    vestopal-W (styrene-unsaturated polyester) 10 ml
    accelerator (initiator: tertiary butyl
        perbenzoate) 0.1 ml
    activator (cobalt naphthenate) 0.05 ml

The activator and accelerator must not be mixed together since an explosion will result. Mix first the polyester and accelerator, and then the activator. The final mixture cannot be kept for more than a few hr.

Infiltration and Embedding: After dehydration the specimens are transferred into 100% acetone and then, if desired, into styrene (100%). Infiltrate the specimens in a solution of 3 parts acetone (anhydrous) or styrene and 1 part vestopal W mixture as given above. After 30 to 60 min at room temperature, transfer into 1:1 acetone (or styrene) and complete vestopal mixture for 30 min and finally into 100% vestopal mixture. All steps vary from 30 min to 1 hr at room temperature and thorough draining of the tissue should be done after each step. Embedding is accomplished after one transfer into 100% plastic mixture in capsules kept at 60° C in an oven overnight.

Removal of Epoxy Plastics 5.4

Sometimes removal of the epoxy plastic from the section is desired in order to examine cellular components such as cellulose microfibrils of a plant cell wall (Fig. 9-10) or bone structure in animal cells. This removal of epoxy plastics may be accomplished by the procedure of Heather *et al.* (1961). Pick up and place the sections on a coated grid. Hold the grids with forceps for up to 30 sec without agitation in a solution of sodium methoxide in benzene (directions for preparing this solution are given below) which has been diluted one third with a 1:1 methanol–benzene solution. Transfer the grids for about 15 sec each through 1:1 methanol–benzene solution twice, 100% acetone, 100% distilled water, and then place on clean filter paper to dry. Shadow the sections with paladium-platinum alloy. The material remaining after removal of the plastic (cellulose microfibrils or bone material) will be separated, and after shadowing, any internal structure or pattern of subunits will be visible on the grid.

The stock solution of sodium methoxide in benzene should be carefully made in a fume hood with no moisture present! Place 2.5 g of metallic sodium in 20 ml of methyl alcohol in a wide mouth flask, and as the solution boils off, carefully add fresh methanol to maintain a 25 ml level. This is a *violent reaction* so the fume hood door should be closed as much as possible. If any moisture is present, there will be an *explosion!!* When dissolved, add 25 ml of benzene and keep adding extra benzene until the solution becomes clear. If a phase boundary continues, add a little methanol. Store the stock solution in a dark glass-stoppered bottle with a large label.

6 BLOCK TRIMMING AND KNIFE MAKING

Block Trimming 6.1

Material embedded in plastic in a capsule mold will require some trimming prior to sectioning with the ultramicrotome. Removal of the capsule can be accomplished by chipping off the gelatin coating or peeling off the polyethylene coat (see Fig. 4-4). Soaking of gelatin capsules does not seem to be too effective for removal of the gelatin. If the rubber molds are not used, the flat embedded material will have to be sawed out of the block with a coping saw so that a rectangle of plastic with the tissue at one end results (Fig. 6-1). The block should be large enough to fit in the microtome chuck. The coarse trimming can be done in a vise while the final trimming should be done under a dissecting microscope with a total magnification of at least 50×. The dissecting microscope should have a calibrated ocular micrometer to permit measurement of the block face.

The block may be pretrimmed with a coarse metal file. The block is held in place by a small table-mounted vise, and the file is used to obtain a truncated pyramid. The tissue is in the center and at the tip of the pyramid (Fig. 6-1). Some investigators recommend the use of a portable dental drill and burr. This technique is very fast, and the results are generally good. However, one must become quite dextrous with the drill or the entire specimen may be removed in trimming. Another minor problem with a dental drill is the large amount of plastic powder that results. A wooden shield built around the region of coarse trimming prevents this powder from coating everything in the laboratory.

Another method for quick and accurate pretrimming utilizes a chamfering mill (Rdzok (1965)). After clamping the capsule in a holder, a $\frac{1}{4}$-in. diameter chamfering tool, which has an angle of 30°, is turned around on top of the block. Usually four to five turns will trim a capsule to a fine point. The equipment can be obtained from most tool suppliers (see Appendix I).

Final trimming is done under a dissecting microscope with a single edged razor blade (Figs. 6-2 and 6-3). The blade should be cleaned in acetone or other solvents to remove any oil placed

6 BLOCK TRIMMING AND KNIFE MAKING

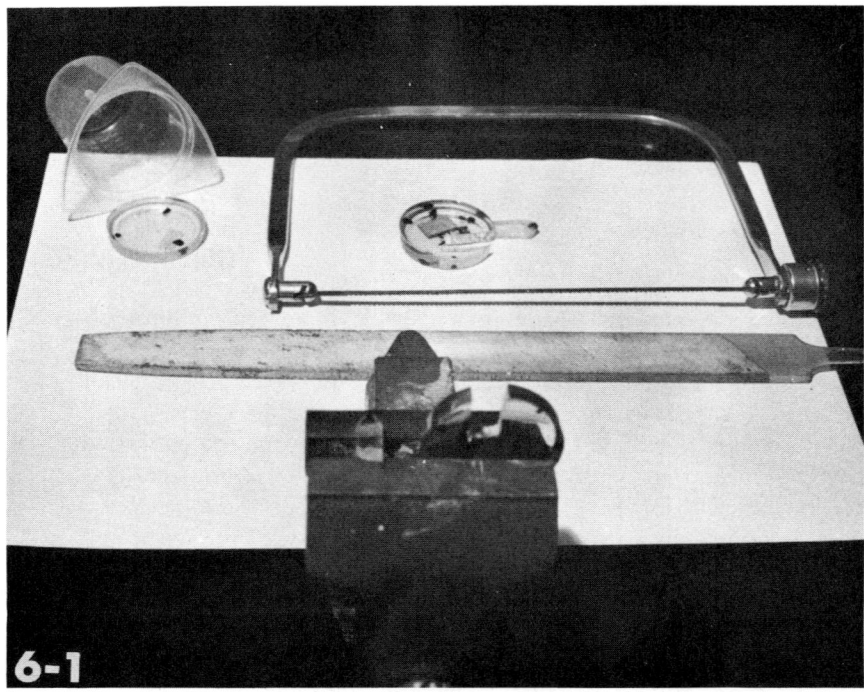

Figure 6-1 Tools needed to cut out flat embedded tissue as described in the text. The material, embedded in a layer of epoxy plastic in a disposable beaker is shaken loose, mounted in a vise, and cut out with a coping saw. A coarse metal file is used to rough-trim the block and to pretaper the block tip.

on it by the manufacturer. The truncated pyramid is transformed into a trapezoid by shaving off alternate sides of the block face until the largest side is less than 0.5 mm in length; this side will be the basal edge of the block during sectioning. The two major sides of the trapezoid should be cleanly cut and parallel to insure a straight ribbon in sectioning. The longest side recommended is 0.1 to 0.2 mm since a large block face is likely to cause chatter

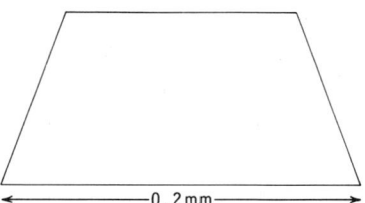

in the sections, and a rapid dulling of the knife edge: The face of the trapezoid should be trimmed down to the specimen. To insure that the tissue is at the face, a thin shaving, made by a razor blade, is examined under the light microscope. Not only will this insure that the tissue has been exposed, but this shaving can also serve as a preliminary orientation for trimming the portion to be sectioned.

In some cases the tissue may require reorientation before sectioning. This is done by trimming off the excess plastic and attaching the block to a wooden dowel with a melted wax solution (25% carnuba wax, 75% paraffin wax). A better procedure is to affix the previously embedded tissue to a block of polymerized plastic by using unpolymerized plastic as glue. Permit the reaffixed block to harden in the oven before trimming. Use of wax as cement is not satisfactory since possible wax contamination of thin sections may occur. Wax melts under the electron beam and contaminates the sections and the electron microscope. The wooden dowel should be of the same diameter as the capsule to insure proper fit in the microtome chuck. Finally, household cement may be used if the block is large enough.

Knives 6.2

A generally accepted explanation of the physical process of ultrathin sectioning is the splitting theory (Sjöstrand (1967)). A section is formed by the surface layer of the plastic block splitting off. The cutting edge of the knife is thought to act as a wedge, similar to the effect of an axe. Thus a small angle between the two facets of the knife edge will aid in producing thin sections. However, a knife edge too narrow in angle (that is, too thin) will be weak and brittle, and it may vibrate causing chatter or thick-thin effects of the section. The theoretical optimum rake angle for ultrathin sections is 30° (Sjöstrand (1967)). Smaller angles can be achieved in metal or diamond knives.

The two major types of knives used today are glass and diamond (Fig. 6-4). Prior to the use of these materials, the only material used was metal. A brief discussion on each of the three types of materials follows.

Metal Knives

Metal knives are not used in ultrathin sectioning currently, although they are commonly used in preparation of sections for light microscopy. Much information on the manufacture and

6 BLOCK TRIMMING AND KNIFE MAKING

sharpening of metal knives is available in the literature. Sjöstrand (1967) describes the usefulness of metal knives prior to the introduction of glass knives, and explains how good quality metal knives can be made and sharpened for ultrathin sectioning. Since sharpening is difficult and tedious, glass and diamond knives have replaced metal knives. One advantage of metal knives is that a rake angle of 30° can easily be made.

Metal knives because of their rapid heat conductivity are useful in sectioning frozen tissues. When frozen tissues are cut, the act of cutting generates heat on the knife which melts a thin layer at the surface of the tissue block. The use of very sharp metal knives prevents this melting and produces quite thin, uniform sections.

Diamond Knives

Diamond knives are commonly used in ultrathin sectioning (Fig. 6-4) because of their convenience of use and long life span. There are a number of manufacturers producing diamond knives with a current base price of about $100 per millimeter of diamond cutting edge. Consequently, a $2\frac{1}{2}$ mm diamond knife mounted in a suitable holder costs at least $350. The useful life span of the diamond knife varies based on actual use and types of plastic and specimens cut; usually the life span is a number of years.

Laboratories with routine research projects which require continuous ultrathin sectioning may have ten or more diamond knives, with a few at the manufacturer for resharpening (minimum cost of $75 per knife). Diamond knives, although expensive, are especially useful in laboratories where the investigator has little or no technical help. The daily routine of glass knife manufacture can be avoided. A further value of diamond knives lies in their extremely sharp and durable edge permitting the cutting of difficult material, such as calcified cell walls. It has been said that epon embedded material will cut well only with diamond knives, although this appears to be more a problem

---

Figure 6-2 The process of trimming a plastic block for sectioning. It is recommended that a calibrated ocular micrometer be installed in the eyepiece of the dissecting microscope. Do not place fingers behind block while trimming, since the razor blade may slip off the block and cut one's finger.

KNIVES 6.2

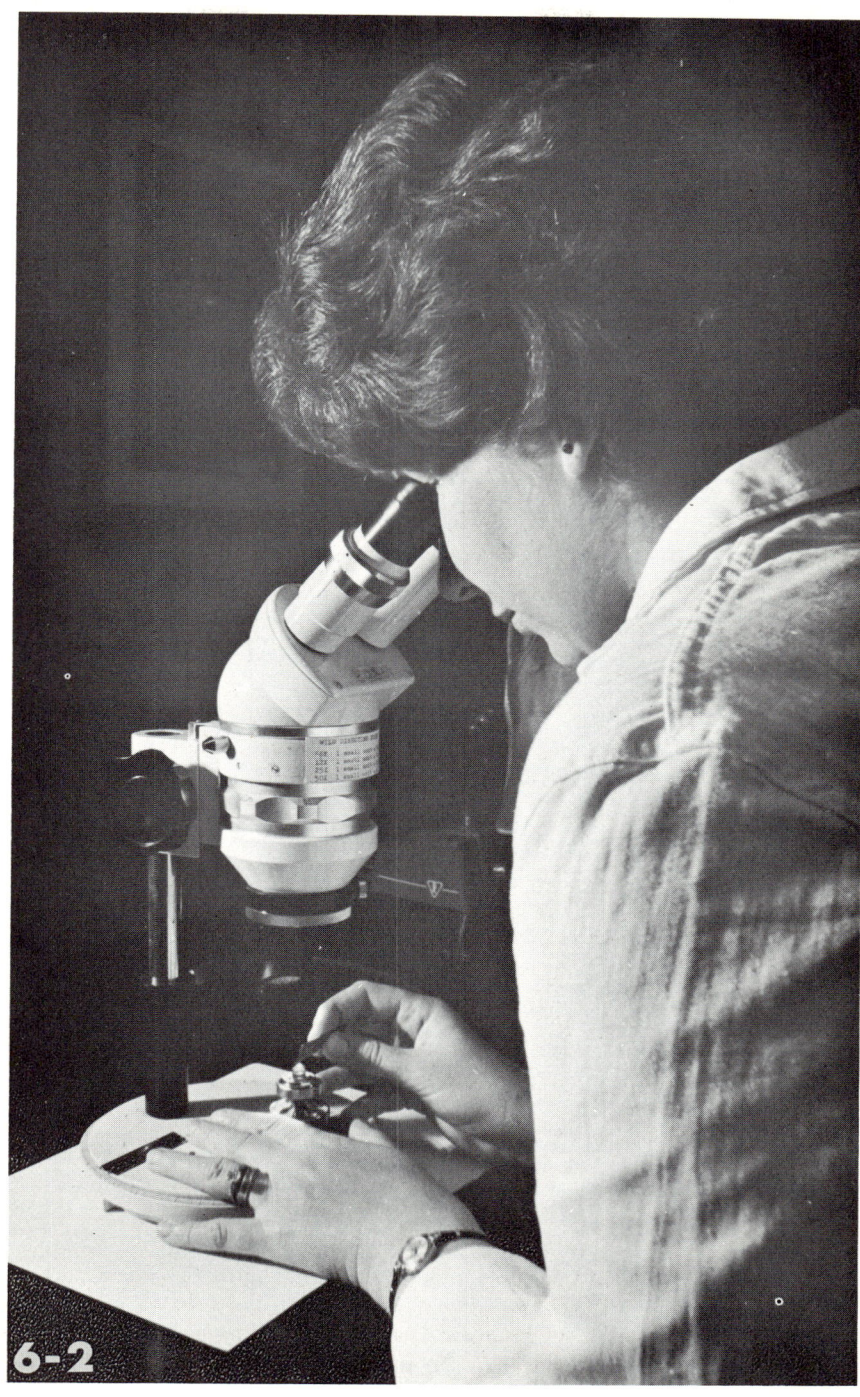

6 BLOCK TRIMMING AND KNIFE MAKING

Figure 6-3 A close-up of a fairly well trimmed block. A cleaned, fresh razor blade is used for the final trimming to insure a smooth, uncontaminated block face.

Figure 6-4 Photograph of three glass knives (items 1,2,3), a metal trough (4), and a diamond knife (5). Knife (1) is shown without a trough, knife (2) has a tape trough made from electrical tape, and knife (3) has a stainless steel trough (4).

KNIVES 6.2

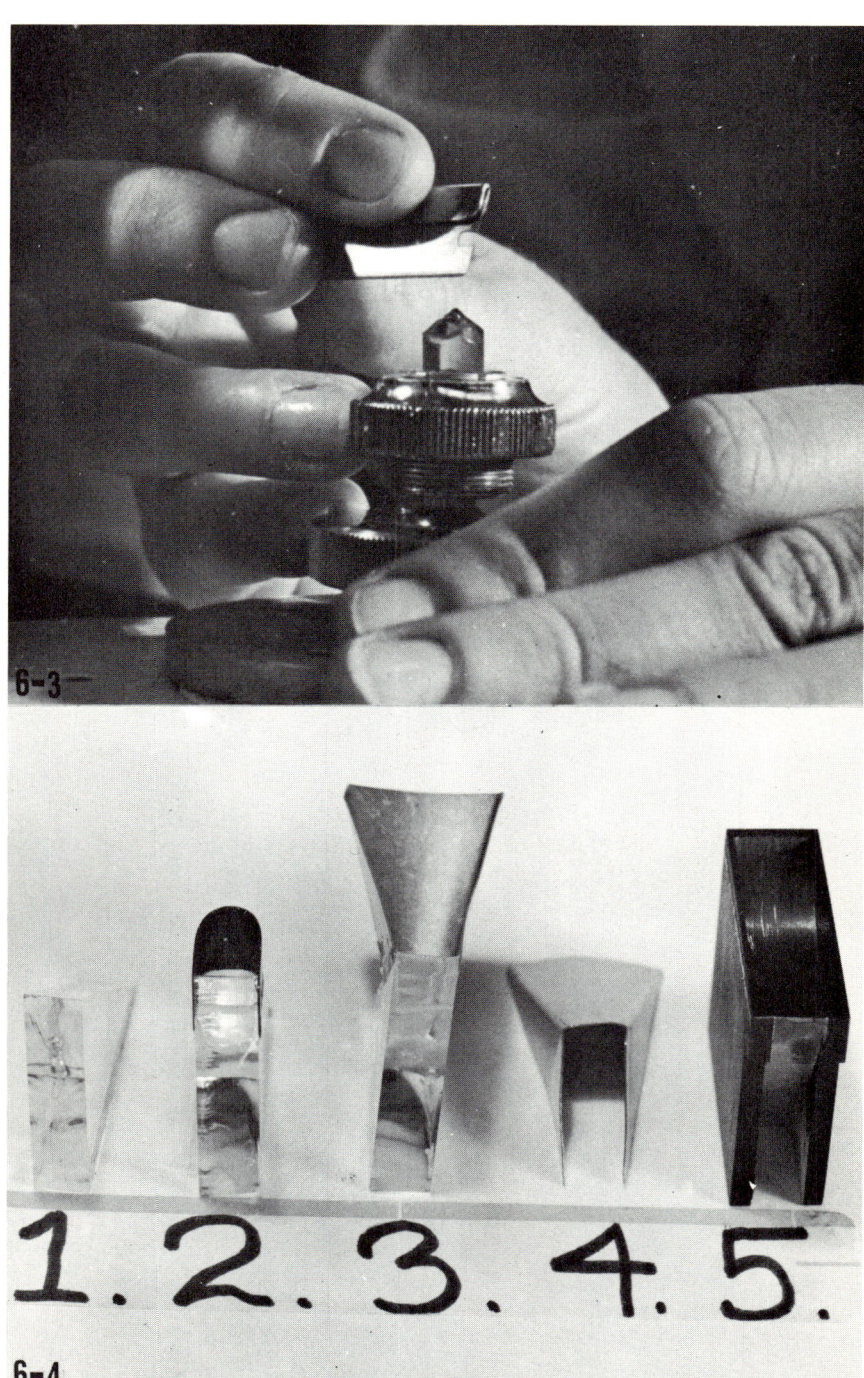

6 BLOCK TRIMMING AND KNIFE MAKING

of good plastic than the knife edge. The disadvantages of diamond knives are their high cost and the constant possibility of damage to these expensive tools.

To prevent material from drying on the knife edge, it is recommended that the knife be stored in a dry and safe container after cleaning. The knife is cleaned with a soft orangewood stick or a flattened toothpick, which is sharpened by a thin chisel edge. The toothpick is run along the entire cutting edge in one smooth stroke that is parallel to the knife edge. If necessary the stick can be soaked in acetone to help remove material sticking to the knife. Finally, the knife should be rinsed in distilled water and dried with compressed Freon or warm air.

Glass Knives

Glass knives are most commonly used for sectioning in electron microscopy (Fig. 6-4) since they are sharp and relatively easy and inexpensive to manufacture. The demand for reproducible results in knife manufacturing has led to the development of knife makers. Two models are presently available (see Appendix I)—the Sunkay Messer Knife Maker (Fig. 6-6, C. W. French Inc., approximately $350) and the LKB Knife Maker (LKB Instruments Inc., approximately $900). Hand manufacture, described below, is also satisfactory. Strain-free glass, which is required for manufacturing knives, can be purchased from various suppliers (Ladd, LKB, or Sorvall). Pyrex glass is also recommended since it appears to be more durable when sectioning epoxy embedded tissues.

Glass knives are made by the free break technique, regardless of the instrument used. The final break, made into a strain-free corner, is a free, nonscored fracture line (Fig. 6-7). The quality, that is the sharpness and flawlessness, of the knife increases the closer the break occurs to the corner. The percentage of flawless edge lies around 90% at a 100 μm distance from the corner, 60% at 150 μm from the corner, 40% at 200 μm from the corner, and varies between 10% and 40% at greater distances. Cross-over knives, that is, knives in which the break occurs at both sides of the corner, are only good if the angle of intersection between the break and the corner is very acute. The usual "roll-over" of the cutting edge, that is, a turning in of the break near the edge to form a more obtuse angle, is also markedly decreased when the knife breaks close to the corner. The reason for this improvement may lie in the fact that the fracture

KNIVES 6.2

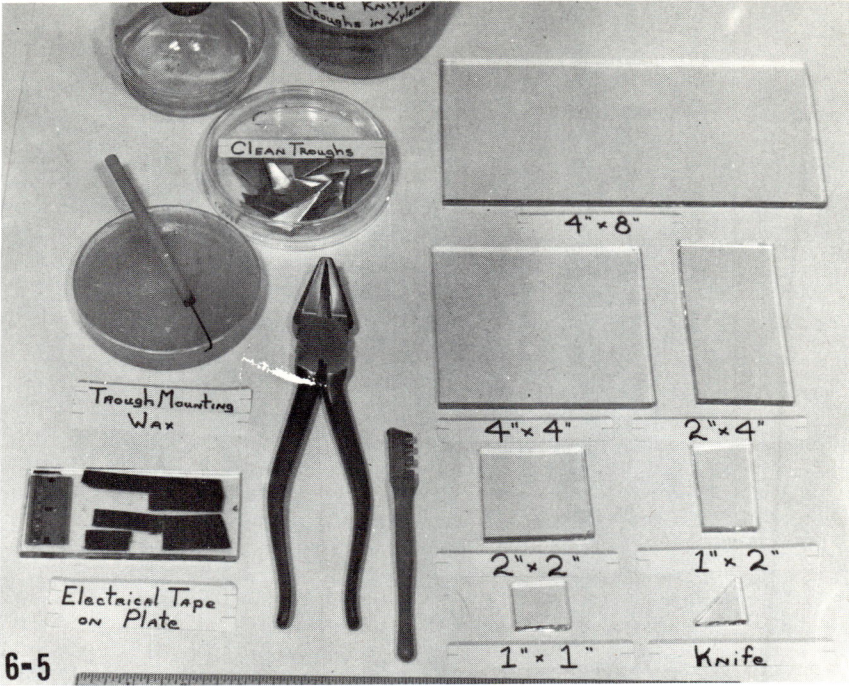

Figure 6-5 Materials needed for the manufacture of glass knives. In the far back row from left to right are an alcohol lamp, a jar containing xylene for used metal troughs, and a petri dish of cleaned metal troughs. The right side shows the recommended series of cut glass plates formed to manufacture a 1-in. glass knife. On the left side are the mounting wax (paraffin/gum mastic), glazier pliers and glass cutter, and a small glass plate with electrical tape and razor blade for tape troughs. (See also Fig. 6-7 for steps in glass knife manufacture.)

separates the two glass pieces only minimally, hence not compressing the future edge before the break takes place.

The standard technique for the manufacture of glass knives by hand employs glazier pliers, hardened wheel cutter, plastic electrical tape or metal troughs, and a ruler (Fig. 6-5). The glass can be bought locally and should be good quality $\frac{1}{4}$-in. plate glass (having a greenish hue) without scratches or score marks measuring 8- by 4-in. The steps given below constitute the pliers technique and are adapted from those given by Ivan Sorvall, Inc. (1965) and others. The glazier pliers and scorer can be purchased locally from a hardware store, or modified glazier pliers are available from Sorvall or Ladd (Appendix I).

6 BLOCK TRIMMING AND KNIFE MAKING

1. Scrub the glass plate vigorously in detergent solution, and rinse in hot water. Stand in a dust-free area to dry, rather than wipe the plate dry to prevent it from acquiring an electrostatic charge which will attract dust particles.
2. Clean the glazier pliers (Fig. 6-5), which should have wide flat jaws about 1 in. in width, with a solvent (acetone). Add three raised pressure points to the insides of these jaws, one at the center of the lower jaw and two at the sides of the upper jaw, to provide the needed forces to stress and break the glass (see Fig. 6-5). Make the raised pressure points out of narrow strips of electrical tape placed over the sides of the upper jaw and center of the lower jaw.

    To break a piece of glass with these modified pliers, line the reference mark on the upper jaw with the score mark (Fig. 6-5). If the pressure applied on the pliers is gradually increased, the break will move slowly across the glass, resulting in a clean break. It is important to keep the jaws of the pliers, especially the pressure points, free of chips of glass so that the pressure will be applied evenly and at the proper places. Align the center tape underneath the score mark on the glass. See the next step to find out how the glass should be cut.
3. Use a good quality glass cutter to make a $\frac{1}{2}$-in. score mark to obtain two 4-inch squares (Figs. 6-5 and 6-7 show the series of breaks described). Kerosene can be used to wet the glass scorer and to give a cleaner score. This also preserves the scorer edge, but it is difficult to clean the kerosene off the cut glass edges. Continue this procedure, keeping the free break opposite the pliers. As Figs. 6-5 and 6-7 show, the series of free breaks will result in a 2×4 in. rectangle, then a 2×2 in. one, followed by a 1×2 in. one, and finally twelve 1×1 in. units of glass. The 1-in. squares should then be scored diagonally, starting 2 mm in from the corner to yield the glass knives.
4. If the free breaks do not yield at least two adjacent sides of each square with clean, smooth sides and no strain lines, the piece should be discarded. The final diagonal break, made with the modified pliers and directed toward the intersection of the two smooth faces will produce one knife from each of the 1-in. square pieces. The closer the break is to this corner of the square, the longer the useful portion of the knife edge will be (Fig. 6-8). Therefore, the piece should be scored diagonally starting 2 mm from this corner. In general, the final break will turn slightly and break out entirely on one side of the corner. The triangular piece on this side will be the knife; the other piece has no cutting edge and should be discarded. Thus, from the original 4×8 in. glass plate, approximately 32 knives can be made.

Figure 6-6 A knife maker (Sunkay Messer) and the series of glass breaks necessary to form a 1-in. knife blade. The large knob on the right side of the knife maker is for increasing the tension of a metal edge directly under the slot seen in the clamping device shown on the left. A diamond scorer is shown on the lower right.

KNIVES 6.2

6-6

6 BLOCK TRIMMING AND KNIFE MAKING

Figure 6-7 A diagramatic view of the process of making 1-in. glass knives from an 8 × 4 in. sheet of glass (see also Fig. 6-5). Free breaks are recommended in all cases. (Diagram modified after Ivan Sorvall, Inc. (1965)).

Figure 6-8 Three examples of glass knife edges showing the three zones of quality of the edge and the shoulder of the break line. Knife (1) is a left-handed knife, while knife (2) is a right-handed knife. Knife (3) is a convex knife with a large region (I) of sharp, mar-free edge.

KNIVES 6.2

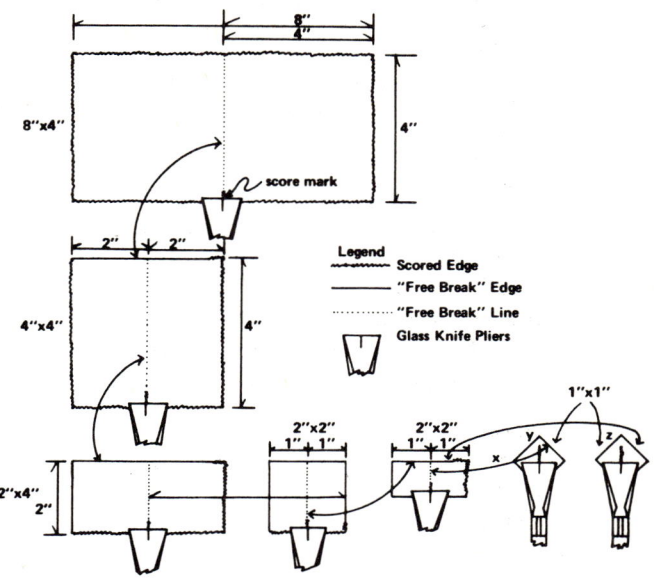

6-7

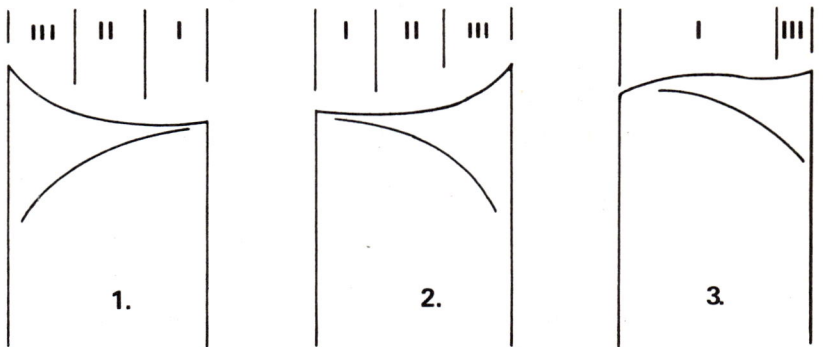

6-8

6 BLOCK TRIMMING AND KNIFE MAKING

5. Control over the location and shape of the knife edge obtained from the final break is best obtained by scoring toward one side of the glass corner or directing the pliers toward one side of the glass corner. If the pliers are tipped (angled) toward the piece that will ultimately be the knife, the knife edge will tend to curve upward to the left. If, on the other hand, the pliers are turned away from the eventual knife, the knife edge will tend to be straighter.

The two types of knives may be identical except that one is a right-handed knife, while the other is a left-handed knife (Fig. 6-8). This can be readily seen by viewing the knife edge at an oblique angle and determining in which direction a major stress line or shoulder of the knife runs. Either configuration might be obtained using the method described above.

Glass knives will vary in the amount of usable cutting edge (Fig. 6-8). The edge may be divided into three zones: I, sharpest, few irregularities; II, fine marks or "whiskers"; and III, unusable.

The size of the peak in region III of each of the knives will increase as the final break moves farther away from the corner of the glass square. Since this portion of the knife is useless, it is desirable to minimize this peak by directing the final break as accurately as possible toward the corner.

Region I indicated on knives 1 and 2 of Fig. 6-8 will be the sharpest and have the fewest irregularities. Irregularities produce knife marks in sections. These irregularities, often called whiskers from their microscopic appearance, are always found in region III and extend more or less across the knife edge toward regions II and I.

Region II may or may not cut well. It is often marred by irregularities, and even when it is smooth it often dulls rapidly. Its use is best confined to trimming in order to spare the good part of the knife edge, region I, for sectioning.

Knives with slightly convex edges (knife 3 in Fig. 6-8) are not often obtained, but are superior in that they are very sharp and the good portion of their cutting edge usually extends most of the way across the knife (region I).

The ultrathin sections cut on a glass knife are floated off on water contained in a trough attached on to the knife, behind the knife edge (Figs. 6-4 and 6-5). Flotation of the sections is necessary because of the delicate nature of the ultrathin sections and the difficulty of picking them up. Troughs can be made from masking tape or metal (Figs. 6-4 and 6-5). Troughs of masking tape offer the advantage of not requiring cleaning since they are discarded with the knife and a new one is used each time, and since they are less apt to leak when accidentally bumped. They can be made by wrapping a strip of tape around the knife and trimming off the excess. They can be sealed around the center edges and coated inside with a melted wax mixture (20% gum mastic and 80% paraffin wax). It is important to coat the inside of the tape with wax to avoid contamination of the trough fluid with the glue on the tape. This is especially true if an acetone–water solution is used.

Metal troughs must be thoroughly cleaned after use (Figs. 6-4 and 6-5). This can be done by soaking the troughs in a 1:1 solution of toluene and xylene, transferring them through three changes of the same solution, next putting them into 95% ethanol, then into distilled water, and finally drying them in the oven. A common problem of metal troughs is wax contaminants which result from insufficient cleaning. These contaminants attach to the sections and make the material worthless for electron microscopy.

The metal trough is attached to the knife by a thin layer of a paraffin and gum mastic mixture (80%/20%). A heated dissecting needle with a bent tip is dipped in a dish of the wax mixture and the wax mixture is spread along the juncture of the glass knife and metal trough. If the wax is hot it will make a good seal around the edge. Care must be taken not to touch the cutting edge of the knife.

7 SECTIONING AND THE ULTRAMICROTOME

Theoretical Considerations 7.1

Ultrathin sectioning of epoxy resins is the process of splitting off of a section from the surface layer of the specimen. To do this the cutting knife acts as a wedge. Consequently, the goal in knife manufacture is to make a very small bevel angle—the angle between the two knife facets (Fig. 7-1). The bevel angle apparently cannot be less than 30°, otherwise the knife edge is weak and will break or vibrate. For glass knives, the optimum bevel angle is around 45°, although some investigators argue that the optimum smallest angle for glass is 50° to 55°. The bevel angle plus the tilt angle of the knife determines the actual cutting angle (Fig. 7-1). The tilt angle permits the specimen to clear the knife after cutting.

The section is apparently produced through a continuous displacement or compression of the surface layer of the plastic block in the direction of cutting. In other words, the section results from the pressure exerted by the knife in the given direction, and therefore the sharpness of the knife edge determines just how well defined the boundary between the compressed and adjacent regions will be. Shearing forces at the boundary produce a crack that advances parallel to the surface.

This explanation is supported by the fact that the amount of compression is directly related to the sharpness of the knife. The above explanation of sectioning is also supported by the fact that the bevel angle for a glass knife can be rather large (50° to 55°) and yet still produce a thin section. Of course with the larger bevel angle, the knife edge is mechanically stable and thus such a brittle material as glass can be used.

Another way of looking at this is to consider the rake angle—the difference between the cutting angle and 90°. If the rake angle is decreased below 30° (so that the cutting angle is greater than 60°), the work on the knife is greatly increased. This work may cause the knife and block to vibrate upon impact. Such a regular vibration causes thick and thin regions to develop in the section, or in the lore of sectioning, chatter. If the excess work on the

7 SECTIONING AND THE ULTRAMICROTOME

knife is too great, the entire section may become distorted or "compressed." These forms of mechanical distortion are explained by the theory of plastic flow; that is, the compression of the block-section at the knife-block interface represents both an irreversible compression and a reversible one (elastic deformation). As the work load on the knife increases, so does the irreversible compression. Too small a rake angle, dull knives, or knicks in the knife all produce this irreversible compression in the section.

7.2 Historical Synopsis of the Ultramicrotome

In this section a brief history of the ultramicrotome is presented. Detailed historical reviews of the development of ultramicrotome and ultrathin sectioning can be found in Pease (1964) and Sjöstrand (1967). Prior to the introduction of reproducible ultrathin sectioning, the primary technique used was whole mount examination, which required fragmentation of the biological material. The plant cell wall, bacteria, and viruses were examined in this manner.

The development of routine techniques for preparing ultrathin sections (about 20 nm) made possible the systematic study of tissue fine structure by about 1952. The starting point for this development began with the microtome being available for light microscopy. The light microscope required sections of not less than 1 nm thick. This represented the minimum thickness that can be achieved by orthogonal cutting (von Ardenne (1939)). Sections thinner than 1 μm broke when bent by the knife. Thus von Ardenne proposed, in order to obtain thinner sections, that they be wedge-shaped and the resulting thin edge of the section be examined.

A number of investigators, including Sjöstrand (1967), attempted to obtain thin sections by the process of double sectioning. Thin sections resulting from a shaved-down block were few in number and difficult to retrieve. An important contribution that led to ultrathin sectioning on a repeatable basis was made by Pease and Baker (1948). They modified a Spencer microtome model 820 so that the tissue block advance mechanism was reduced to $\frac{1}{10}$. They also used the double embedding media of collodion-paraffin and reduced the block face to 1-mm square. Sections of 0.3 to 0.5 μm were then obtained.

Further modifications of the Spencer microtome model 820 by Hillier and Gettner (1950) resulted in sections 0.2 μm thick.

HISTORY 7.2

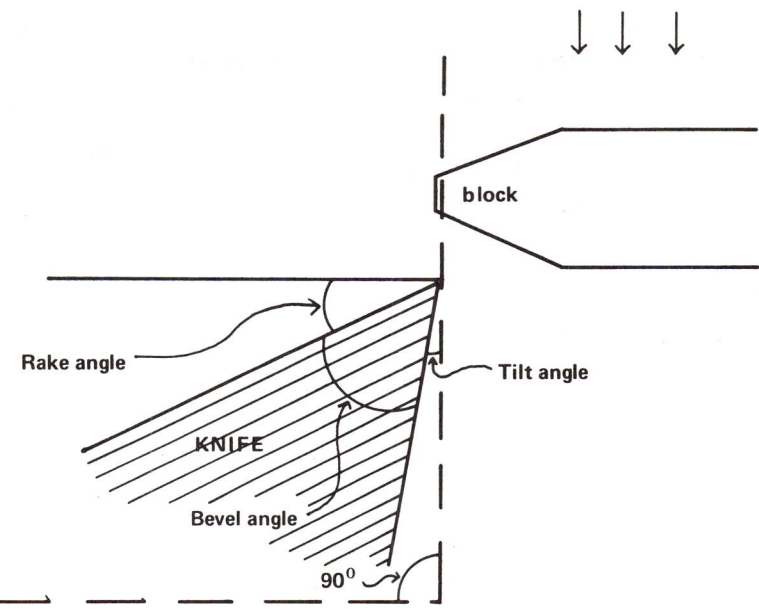

Figure 7-1 The knife-block relationship showing the various angles discussed in the text. The "cutting angle" of such an arrangement is the tilt angle plus the bevel angle of the knife.

They removed some of the friction in the advancement of the specimen, improved the sharpening techniques for metal knives, and used a trough on the metal knife to float off the sections. Hillier also recommended against removal of the embedding media, a common step in preparation of light microscope sections.

Glass knives were introduced by Latta and Hartman (1950), and plastic embedding had already been introduced by Newman et al. (1949). In 1952, Palade working in Porter's laboratory used methacrylate embedding and glass knives on a microtome designed by Claude and Blum (Claude (1948)). He was able to obtain sections 100 nm thick.

Porter's work on an improved microtome made possible the studies of Palade. This microtome used a metal bar with one end fixed and the other end movable by bending the bar, thus bypassing the cutting edge on the return stroke. The free end of the bar formed a parallelogram in movement, while the fixed end was supported in a double set of pivots. These improvements are the basis for the Sorvall Porter–Blum Ultramicrotome.

Sjöstrand, in 1952, showed high resolution electron micrographs of material sectioned by using a highly sharpened razor blade and a specimen block face of less than 1 mm^2. In 1953, Sjöstrand reported a new type of microtome that used an oil film on a heavy precision bearing with a well defined rotor. Coupling this with other improvements (sapphire bearings, knife retraction, and thermoadvance) the Swedish firm, LKB Instruments Inc., developed the LKB Ultramicrotome.

7.3 Current Ultramicrotome Models

Presently, there are a number of models of ultramicrotomes available with prices ranging from about $1500 to over $6000. A major factor in cost is the ease of operation. Specimen advance is either thermal or mechanical. The specimen movement is either circular, in a parallelogram, or strictly up and down with knife or specimen retraction on the return stroke. Since it is not the purpose of this text to compare and discuss individual ultramicrotomes, only the Sorvall MT-1 will be used here as an example. A list of major models and companies is given in Appendix I.

Any ultramicrotome can be divided into the knife-holding and specimen-holding assemblies, advance system, and observation-illumination system. The Sorvall MT-1, which has been in existence since 1953, is a good instrument to refer to since mechanically it is a simple device and is well understood.

The advance system of the MT-1 is mechanical (as opposed to thermal advance). The specimen, mounted on a cantilever arm, is advanced by the turning of a lead screw. The amount of advancement can be regulated by a cam device which controls the amount of turning of the lead screw per cutting stroke. The increments of advance in the MT-1 are 25 nm and are dependent on the size of the screw threads. A thermal advance instrument (LKB, JUM-5, Reichart, or Leitz) uses a controlled source of heat to expand the specimen arm rather than a mechanical lead-screw arrangement. The actual cycling of the specimen in sectioning is done by turning a flywheel by hand in the MT-1, while other instruments may have motor drives.

The specimen–knife-holding assembly varies in complexity and adaptiveness according to the instrument. The Sorvall MT-1 knife holder is relatively simple. The knife, with attached trough, is clamped into the holder by a Teflon tipped screw. The entire assembly can be moved back and forth on a stage track, and the knife holder can be pulled in and out to allow proper positioning of the knife with relation to the block. Tilt of the knife is accomplished by adjusting the knife holder unit. No other movements of the knife holder assembly are possible. Both coarse and fine adjustment (micrometer) knobs control the advancement of the upper portion of the knife holder assembly. The specimen is mounted in a ball-bearing chuck which is attached to the advancement rod. This type of attachment permits orientation of the block around 360°. More complex instruments (Sorvall MT-2, LKB Ultratome III) also permit circular rotation of the knife holder assembly allowing greater flexibility in specimen-knife orientation.

The observation-illumination system is a modified dissecting microscope system with a cold lamp source (to prevent thermal expansion of the plastic block or specimen arm). The Sorvall MT-1 uses a swing arm mount for the dissecting microscope which allows tilt, lateral, and in-out movements for positioning of the dissecting microscope. The illumination system is a cold fluorescent lamp which also can be moved in and out and in a semicircle for proper orientation on the knife and specimen. An ocular micrometer may be placed in the dissecting microscope for measurement of section size and ribbon length.

Modifications of the Sorvall MT-1 are described by Pease (1964) and similar modifications have been carried out in our laboratory (Fig. 7-2). By reducing the cutting speed in half through the addition of two gears with a ratio of 2 to 1, a much more steady cutting stroke is produced (Fig. 7-2). Instead of gears, however, small "v" belts can be used which nullifies the possibility of gear vibration. Another helpful, minor modification of the MT-1 is the strengthening of the fine and coarse advancement bar of the knife holder assembly. A hole can be drilled through the cantilever arm of the advance system on the knife holder assembly and a brass screw placed in it.

Sectioning 7.4

The art of sectioning requires practice, especially when the less expensive hand driven ultramicrotomes are used. Since each instrument has its own distinctive procedure for sectioning, only some practical steps will be outlined here. The reader should

7 SECTIONING AND THE ULTRAMICROTOME

remember above all that good, useful sections and consistent ribbons of sections require a uniform and patient procedure.

In setting up the ultramicrotome for sectioning, two general concepts must be kept in mind—microtome stability and air movements. The microtome should be placed on a sturdy bench or table. A table can be built from two stacks of four concrete building blocks for the legs and two 42-in. long concrete lintels as the table top. These are inexpensive, easy to obtain, and produce a most stable table (Fig. 7-2) if newspaper is placed between the blocks to prevent rocking. The microtome may be placed directly on top of the lintel. The table should not be placed in the path of air currents. A homemade Plexiglas shield may be placed around the cutting unit (knife holder assembly). Materials that should be available (Fig. 7-3) during sectioning are:

1. a squeeze bottle and 10-ml syringe and needle containing distilled water (or 10% acetone) for filling the trough;
2. a dropper bottle of distilled water to keep the trough at the proper level;
3. clean grids, coated or uncoated (see chapter 8);
4. a slice of filter paper about 10 cm long;
5. a clean razor blade and toothpicks;
6. a dropper bottle of toluene or chloroform for spreading the sections;
7. a container of tissues for cleaning off the block face and drawing off excess liquid from the trough;
8. a good pair of sharp forceps; and
9. an eyelash hair with the thick end taped to an applicator stick.

The knife with attached trough should be firmly mounted in the holder. The knife edge should be flush with the front of the holder. The clearance or tilt angle (see Fig. 7-1) should be about 2° to 5°. Positioning the knife properly should insure that the

Figure 7-2 A Sorvall MT-1 ultramicrotome mounted on a pair of cement lintels set on two stacks of four concrete building blocks, with newspaper interleafed to prevent rocking. A slab of $\frac{1}{4}$-in. hard sheet rubber is placed on top of the two stacks of blocks (arrow) and the two concrete lintels are laid on top. The lintels may then be covered with self-adhering vinyl paper for esthetic reasons. The MT-1 has been modified by reduction of the cycling ratio to $\frac{1}{2}$ for each turn of the handle. This insures a slower movement during the cutting stroke and decreases the effect of direct motion of the hand on the cycle.

SECTIONING 7.4

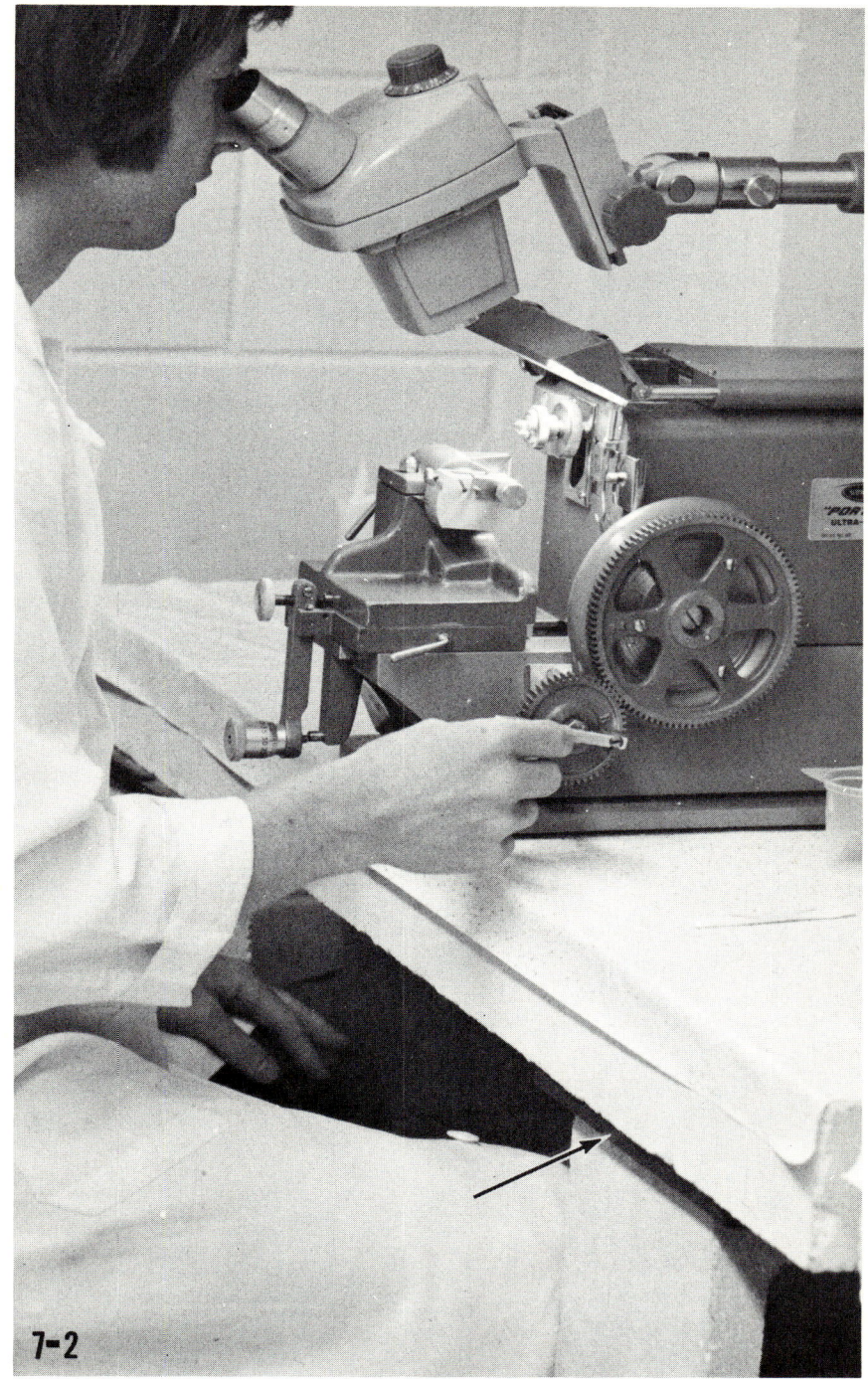

7-2

7 SECTIONING AND THE ULTRAMICROTOME

specimen will be cut by a good region of the knife. Use a less desirable region (region II) on the knife for initial block face trimming and then move the knife to where the edge is good (region I). For preliminary orientation the knife should be set back slightly from the specimen block. The knife tilt can be checked from the side, if no degree of tilt scale is given on the holder.

The specimen block is firmly mounted in the chuck or specimen holder with the long side of the trapezoid down and parallel with the knife edge (level). In addition, the face of the block should be flush (smooth) with the knife facet. If not, use a poorer region of the knife (region II) and face the block by advancing the knife by hand and cycling the block. When this is completed, clean off any shavings of plastic and recenter the knife.

The block face can be cleaned of dust or contamination (drops of water) by gently passing a strip of lens paper over it. During adjustments, the dissecting microscope should be focused on the knife block region and the illumination oriented so that the knife edge is lighted.

The actual process of sectioning may now begin. Recheck to insure that all of the knife and block mounts are firmly set. Set the advance mechanism for about 100 nm and begin cycling the specimen. Gradually move the knife holder assembly forward with the coarse and then fine adjustments. When the knife edge appears to be just about to cut a section, carefully advance the knife holder assembly at about 1 μm per cutting stroke. After one becomes familiar with these steps, judgment as to when the knife and block are very close will become more accurate and little time will be spent in the preliminary advances of the knife. As soon as a few thick (less than 1 μm) sections are made, cease advancing and lock the knife holder assembly in place. Now fill the trough with either distilled water or 10% aqueous acetone. Distilled water is usually preferred because acetone tends to carry in solution a number of common contaminants (trough tape adhesive, wax). The meniscus should be low (slightly negative) to insure adequate refraction of light from the sections. Finally reexamine the knife edge to insure a good clean edge and a clear solution in the trough.

The process of sectioning should utilize a smooth steady stroke, when done by hand. The speed during the cutting portion of the cycle is most important. Too fast a speed will result in chatter on the sections, while too slow a speed will usually result in jerky movements of the hand. A good speed will permit the operator to follow the block during the entire cutting stroke. There is a rhythm to cutting and only practice will perfect it. Interruptions from other persons, phone calls, or other demands in the laboratory should be avoided when sectioning. Do not hold the cutting hand wheel too firmly otherwise vibrations may occur and chatter will ruin the sections.

Once some sections start coming off, reset the advance mechanism so that pale gold (straw colored) sections will be produced. Gradually change the setting so that silver sections will dominate the ribbon. Make certain that the ribbon is straight and that no knife marks are visible on the sections. Retrim the trapezoid or find a better region of the knife if these problems occur. If chatter (thick and thin regions appearing on each section) is obvious, slow down the cutting speed and check that all the units are tightly sealed. The knife edge can be cleaned with a toothpick which has been shaved to a thin edge. Dirt or scum can be removed from the trough liquid by dragging a lens tissue over the surface and replacing the liquid.

The thickness of the sections can be determined by the interference colors produced when reflected light is viewed through a dissecting microscope (Peachy (1958)). The meniscus in the trough may have to be lowered or raised to see the interference colors of the sections. According to Peachy (1958) the following section thicknesses will produce the following characteristic interference colors:

60 nm	gray
60–90 nm	silver
90–150 nm	gold
150–190 nm	purple
190–240 nm	blue
240–280 nm	green
280–320 nm	yellow

Most investigators tend to select the thicker sections (70 to 100 nm), and thus those with greater contrast, when using the electron microscope at low magnifications (10,000 to 30,000$\times$). When using the electron microscope at higher magnifications (50,000 to 100,000$\times$), investigators will tend to select the thinner, high resolution sections of 25 to 50 nm. A beginning investigator should aim for ribbons of pale gold to silver, or thicknesses of 100 to 70 nm, and then an attempt should be made to obtain thinner sections.

7 SECTIONING AND THE ULTRAMICROTOME

Figure 7-3 The materials that should be available for ultrathin sectioning. The distilled water in the squeeze dropper bottle and the syringe are for filling the trough. Toluene is used on the wand (a folded piece of paper) to cause the sections to spread out. Toothpicks are used for cleaning off dirty glass-knife edges and for transferring thick sections to glass slides.

Figure 7-4 The process of picking up a ribbon of sections floating on the water in the trough. The block is stopped below the knife, and the sections are spread out by waving a wand with toluene over the sections. The grid, held at the very edge by a pair of finely pointed tweezers, is brought down onto the ribbon. Care must be exercised not to touch the knife's edge with the grid or forceps.

SECTIONING 7.4

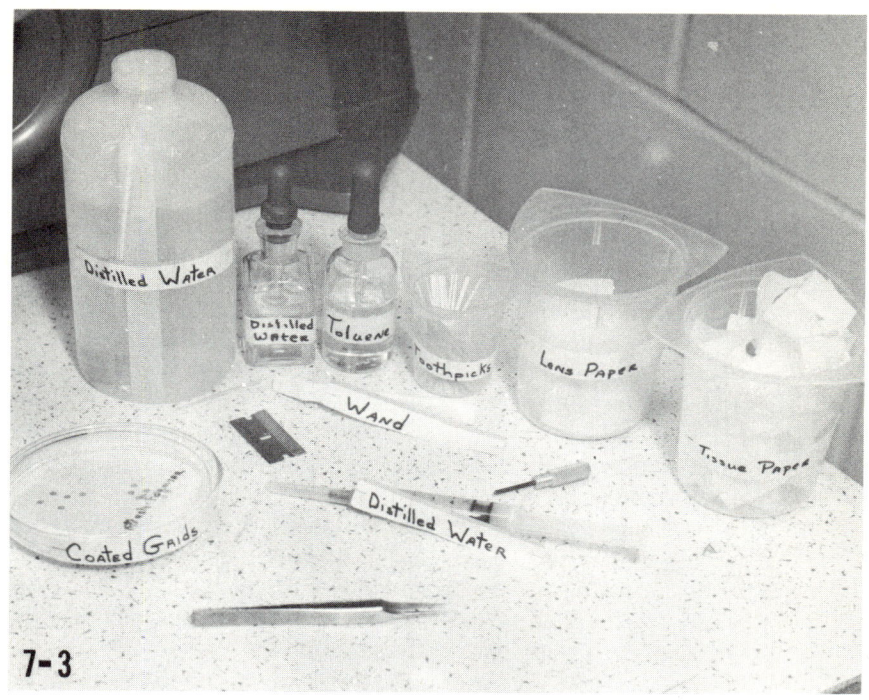

7-3

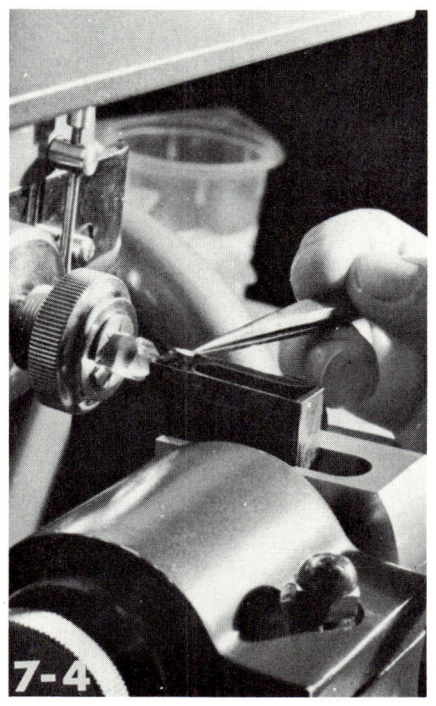

7-4

103

7 SECTIONING AND THE ULTRAMICROTOME

When a ribbon of sections has been cut, the operator may wish to expand the sections to remove any folds or wrinkles. Most epoxy resins are rather inert to the vapors of solvents and do not expand well, although some spreading can be accomplished with toluene or chloroform vapors. Heat, in the form of a small bulb connected to a rheostat, will also expand epoxy sections. The methacrylates, on the other hand, expand readily with chloroform. To expand the sections with solvents, place a drop of solvent on a thin strip of #1 filter paper and gently wave the paper over the sections. Care should be taken so that the solvent-soaked filter paper tip does not touch the liquid or sections in the trough.

The sections can be easily picked up by touching a clean, dry grid to the ribbon as it floats on the liquid (Fig. 7-4). Since the epoxy or methacrylate sections are hydrophobic they will remain attached to the grid surface. The grid screen must be dry and may be coated with a plastic film (see chapter 8) or left naked. An alternative method is to bring the grid up under the floating sections. Although this procedure requires greater dexterity, it usually prevents folding of the sections during pickup. After picking up the sections, place the grid on a fine, fiber-free filter paper to dry, with the section side up.

The filter paper may be placed in the petri dish with the identifying information written on the paper. All grids containing sections cut from the same block are placed in sequence on the paper with the remaining part of the block after it has been sectioned. Fine jeweler's forceps with pointed tips are used in handling the grids. The grid is carried by a small lip which is bent back for gripping by the forcep tips.

The technique described for picking up the sections is fast and simple, but there is always the danger of causing folding of the sections and damage to the knife edge (especially diamond knives). To avoid touching the knife edge, the ribbon can be teased away from the edge with a dry toothpick or eyelash (tapered side) taped to an applicator stick. After the sections are floating free from the knife, they can be picked up.

Another technique is to transfer the entire ribbon of sections from the trough via a small wedge of a glass. A broken glass slide makes a most effective glass spatula. This latter technique requires some dexterity since the wetted edge of glass (or spatula) is pushed underneath the ribbon from behind, and the

SECTIONING 7.4

sections tend to migrate up the wedge (or spatula) due to surface tension. After removal of the ribbon from the trough the grid can be touched to the ribbon.

Serial sectioning is no different from normal sectioning, except that every section produced is needed and must be of even thickness. To do this requires some practice and a reliable ultramicrotome. The main problem is to acquire a collection of uniform ribbons of sections so that a continuous sequence results with only a minimum of unusable sections, that is, lost or folded, or found lying on a grid mesh. Slotted grids coated with a plastic film are recommended (see chapter 8). The grid is maneuvered so that the ribbon lies directly under the slot, and then this grid is touched down on the ribbon.

The following outline summarizes problems encountered in ultrathin sectioning and gives possible causes.

1. *Skipping (or only obtaining) sections on every other stroke:* Knife is tilted too far back and therefore the block hits the knife face; the block is too large; or the block is loose.
2. *Chatter or thick-thin regions (bands) across each section:* Knife tilt is too great causing too small a rake angle; knife is dull; knife is loose; block is too hard; or the cutting was too fast or performed with an unsteady stroke. If chatter is only found in local regions of the section, then the cause may be local dull spots in the knife or poorly embedded regions of the specimen.
3. *Specimen picked up by passing block:* Plastic is not hard enough; knife or block faces are wet; or the specimen or block face is dirty.
4. *Knife marks or streaks appearing on sections:* The knife edge is bad or dirty; or the crystalline material in the specimen causes tears.
5. *Compression or general contraction of the section:* Knife is dull; plastic is too soft; or the specimen block is loose.
6. *Thickness varying from one section to another:* Block is loose; plastic is soft; clearance angle is too small; or the air movement in the room causes contraction or expansion of the plastic block.
7. *Folding of sections at knife edge:* Trough liquid is too low (knife edge dry); or the plastic is soft.
8. *Contamination or dirty sections:* Reagents (water or acetone) are contaminated; trough is not cleaned properly (xylene remains, see chapter 6); knife is dirty; or the block face is dirty.
9. *Grainy to poor sections:* Knife is dull; tissue is not well embedded; or plastic is not completely polymerized.
10. *No sections at all:* Air movements or block temperature cause contraction; knife is too far back so it does not touch the block face; or the microtome advancement is used up.

8 PREPARATION OF SPECIMEN GRIDS

Types of Grids 8.1

The common procedure for picking up sections for viewing with the electron microscope requires specimen screens or grids. These discs are either 2.3 or 3.0 mm in diameter, depending on the make of the electron microscope, and can be purchased from general electron microscope supply houses (see Appendix I). The specimen screens can be made from a variety of metals (copper, nickel, gold, paladium, or molybdenum); copper grids are the most common. The screens are made by one of two processes—punching out a sheet of screening or photoengraving and etching each individual grid. The punched screens tend to be rather rough around the edges because of the punch technique used and the mesh is somewhat uneven because of the electrolytic process of deposition. However, for most purposes punched screens are adequate and are about $\frac{1}{4}$ the cost of photoengraved grids (Athene type). The more expensive Athene, or Athene-type, grids are smooth on both sides, have a border, and usually have a shiny side and a matte, dull side.

The specimen screens, especially the Athene type, come in a variety of mesh sizes and special forms. The standard mesh sizes of 75, 100, 200, 300, and 400 refer to the number of wires per inch, and therefore these sizes indicate the amount of open area of each mesh size (Fig. 8-1). For example, a 100-mesh grid should have 100 bars (or wires) per inch. Although the actual amount of open area in a given mesh size varies from one manufacturer to another, a 75-mesh grid has about 75% open area, a 100-mesh about 65%, a 200-mesh about 45%, a 300-mesh about 40%, and a 400-mesh approximately the same. Variations in mesh size can produce unequal mesh type grids (75 × 100-mesh; 100 × 300-mesh). Special slot, hole, or bar grids are also available and are used in serial sectioning.

The use of modern epoxy plastics has reduced the need for film-coated grids which were necessary in the era of methacrylate plastics. Standard sectioning and observation usually require only

8 PREPARATION OF SPECIMEN GRIDS

clean, naked grids. However, coating films are required for high resolution microscopy, grids with larger mesh openings (75- and 100-mesh), and slot/bar grids.

8.2 Grid Cleaning

Before applying the supporting film, the grids should be cleaned. Very dirty grids may be soaked in detergent, rinsed in water several times, rinsed in a reagent grade acetone, and allowed to dry on filter paper. However, most of the time, simple degreasing by rinsing in acetone and drying on filter paper is all that is necessary. Reuse of grids can be tried, but this is not recommended because any cleaning procedure will be only partially successful.

The following procedure is used successfully in the author's laboratory to clean grids. The grids are placed in 95% acetone for 5 sec, 0.1N HCl for 5 sec, distilled water for 5 sec, 0.25% formvar in ethylene dichloride (see below) for 1 sec, and dried on filter paper. The final plastic coat gives the grid a hydrophobic coat to which sections or plastic film will adhere. This last step must be performed rapidly to avoid strands of plastic forming across the openings. Many laboratories store cleaned grids in 100% acetone, drying them on filter paper just before use.

8.3 Supporting Films—Plastic

It is possible to use naked grids and it may even be preferred. However, unless a section covers the entire grid hole and itself is free of holes and scratches, it may drift or creep under the beam making photography impossible. This difficulty can be overcome to some extent by using grids with small openings (e.g., 300-mesh), by stabilizing the sections by evaporating a coat of carbon on them, or following both procedures.

Usually, however, a thin film is applied to the grid to support the sections across the square holes. This film reduces both contrast and resolution in proportion to its thickness. Pure carbon films can be made by deposition of vaporized carbon *in vacuo*. They are thin and very stable, but they are difficult to make, stick to the grids poorly, and are brittle. Plastic films (parlodion or formvar) do not have these disadvantages, but they tend to creep or drift under the electron beam. Such movement tries the patience of microscopists attempting to obtain micrographs at high magnification.

Parlodion is nitro cellulose (collodion) and is available from Mallinckrodt Chemical Works, St. Louis. To make a stock solution of 0.5% (or less) to 2%, the parlodion is cut into small pieces and placed in amyl acetate. The parlodion swells and dissolves slowly; two days with occasional stirring may be necessary for complete solution.

Formvar is a polyvinyl formol plastic available from Shawinigan Products Co., New York. Stock solutions of 0.2–0.5% in ethylene dichloride can be made from powdered formvar. Pease (1964) has suggested that solutions less than two days old give better quality coating films. Three methods are described for film manufacture.

Method I A 0.5% (or less) solution of parlodion in amyl acetate or a 0.25% solution of formvar in ethylene dichloride is used in a wide-mouth bottle or a Coplin jar. Concentrations can be varied to obtain the desired thickness of the film. Different batches of formvar, however, may require different concentrations to obtain the same thicknesses of film (Fig. 8-2).

A clean (but not too carefully cleaned) 1- by 3-in. glass microscope slide is dipped into the solution. The excess drops are rapidly drained by touching the lower edge of the slide to filter paper, and the slide is left to dry (in a vertical position) in a dust-free place. When the film is dry, score the edges of the slide around the area of the film.

Next, float the film off onto a clean water surface provided by a glass bowl filled with very clean distilled water. To do this, the slide is "frosted" by breathing on its surface which coats it with condensed water vapor and, while still frosted, the slide is dipped into the water at an angle of about 30° (Fig. 8-3). The film should peel off and float onto the water surface as the slide sinks down. The grids are laid upon the floating film with their smooth surfaces touching the plastic film. To assure a good bond between the grid screen and the film, each screen is gently tapped with a needle so that the film is depressed. The film with its grid is picked up from the water surface by touching the film with a piece of paper and carefully inverting the paper-film. A dry glass slide can also be used to pick up the film. The slide is brought down on top of the film (Fig. 8-4) and then underwater where the slide is inverted and the film and grids, now on top of the slide, are brought out of the water for drying. After the grids are

8 PREPARATION OF SPECIMEN GRIDS

Figure 8-1 Four common Athene-type grids with different mesh sizes. Grids of 200-mesh, or larger openings, should be coated with a plastic film if a stable section is desired.

Figure 8-2 A demonstration of grid coating. In the foreground on the left side are coated slides leaning against corks to dry. Two forms of pickup screens are seen behind the coated slides and these can be used to pick up grids with a plastic film on top. Two small wide-mouth jars containing formvar (0.25%) in ethylene dichloride and parlodion (0.5%) in amyl acetate are located behind some coated grids mounted on glass slides in a petri dish. The large finger bowl to the right of this demonstration is filled with distilled water and its surface is used to float films off.

PLASTIC FILMS 8.3

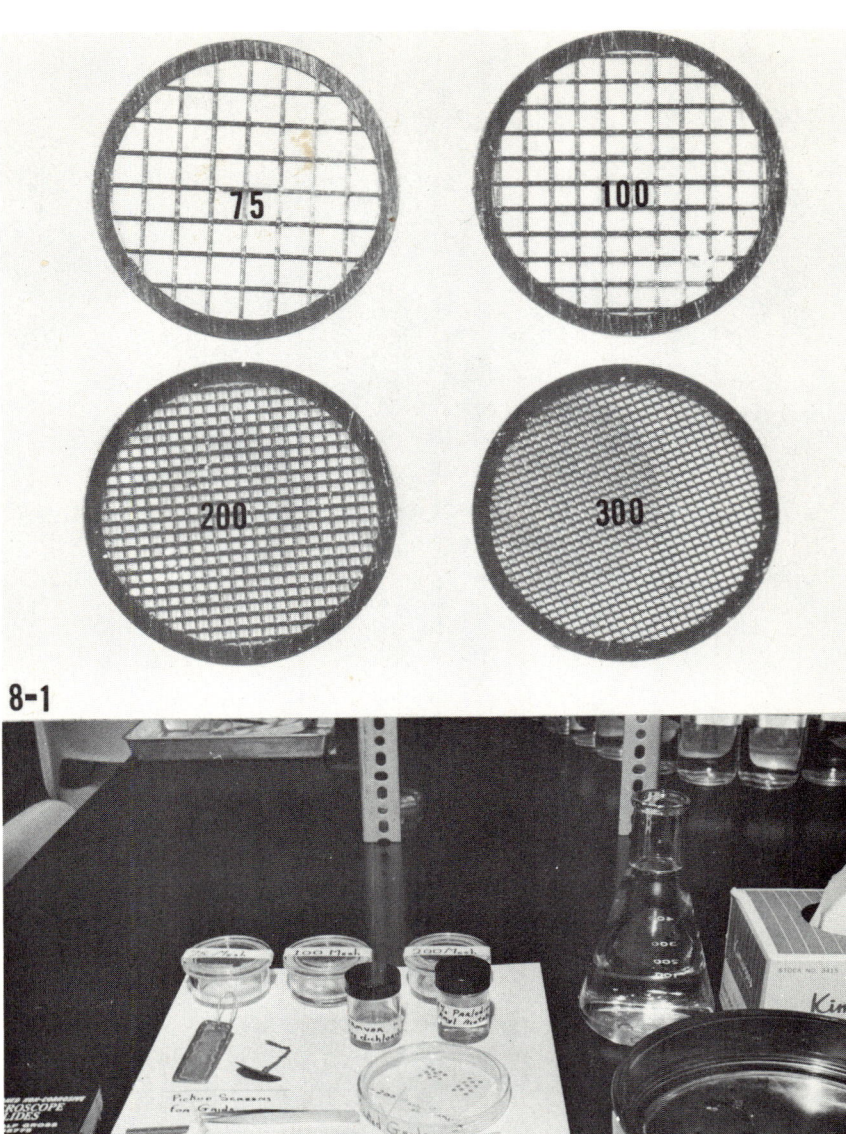

8-1

8-2

8 PREPARATION OF SPECIMEN GRIDS

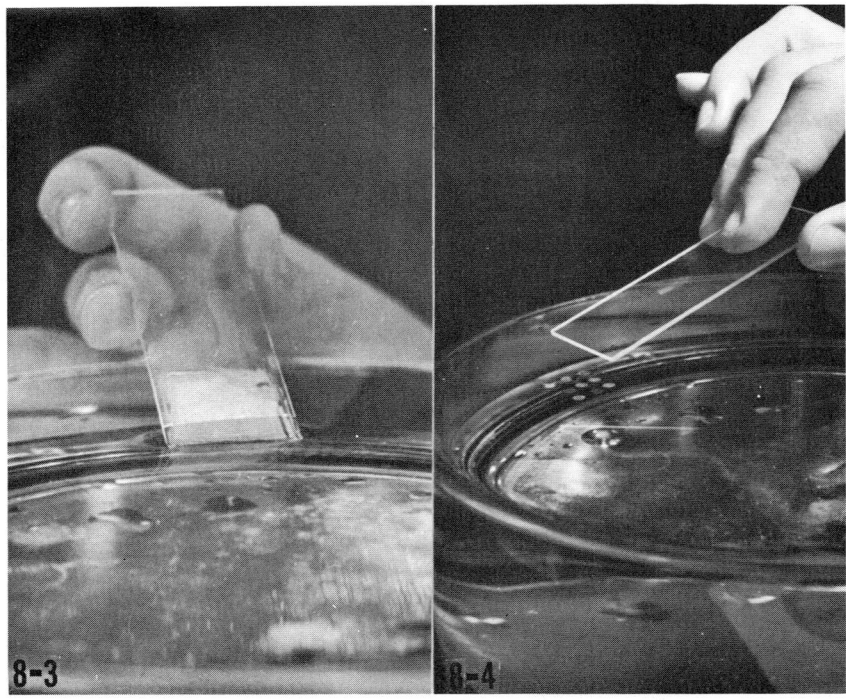

Figure 8-3 After the plastic film on the slide has been frosted and the slide has been "scored" to permit film removal, the slide is touched to the surface of the water at an angle of 45°. The film floats off onto the water.

Figure 8-4 The grids, now placed face down on the plastic film floating on the water, are picked up as shown. A glass slide is placed over the grids on the plastic film and brought straight down into the water. The slide is then turned so that the grid side is up, and the slide is lifted out of the water. Figs. 8-2, 8-3, and 8-4 are all demonstrations of Method I for the process used in manufacturing a coating film.

dry, a coat of carbon can be evaporated on them, or they may be used without the stability offered by the carbon.

Method II A film can also be made by dropping one or two drops of 0.25% formvar or 1–2% parlodion solution onto a large, clean water surface. Surface tension spreads the solution on the water before evaporation sets the film. A 3- by 4-in. staining dish or a circular bowl $3\frac{1}{2}$- to 4-in. in diameter can be used. The grids are placed on the film as with Method I, picked up by laying a strip of brown wrapping paper on the film, and then coated with evaporated carbon in a high vacuum evaporator.

Method III A 6-in. circular container is set in a dish and filled to overflowing with distilled water. Then a 1- by 3-in. strip of screen (20-mesh to 60-mesh) is placed under the water but some distance from the bottom, on an upturned small beaker. Grids are then arranged on the screen with the side to be coated facing up. A film is next cast on the water surface as described in Method II or floated off a slide as described in Method I. In the case of Method II, film wrinkles can be avoided by covering the dish as the solvent evaporates. Now, with the film floating free, the water is slowly removed from the container by either draining through a spout at the base or simply siphoning off through a small diameter hose. This allows the film to lower onto the grids. When dry, the grids can be covered with a carbon layer to strengthen them or left as is. Alternatively, one may carefully pick up the screen holding the grids and the floating film. However, this method may result in wrinkles.

Preparing films with holes Films with holes are prepared for the purpose of electron microscope lens correction to eliminate astigmatism and to estimate the microscope's resolution.

A glass slide is dipped into a plastic solution as in Method I. Immediately after withdrawal and before the film dries, the slide is held over steam or is frosted by breathing on it.

The number and size of the holes formed is controlled by varying the length of time the film is exposed to the steam, or by partly drying the plastic over steam and then allowing it to dry in a container with a high humidity. As in Method I above, the film is removed from the slide, grids are placed on it, and it is picked up from the water surface. After drying, the grids are coated with carbon and are ready for use. It is possible to enlarge the holes if they are too small by careful exposure to the electron beam in an electron microscope before the evaporated carbon coat is applied. If desired, a carbon film with holes can be made by removing the formvar. This is done by placing the grids in a vapor of ethylene dichloride for several hours.

Supporting Films—Carbon 8.4

Carbon films are strong, beam stable, and can be made quite thin. Carbon films scatter electrons only slightly, thus they do not offer resolution problems. They are brittle, however, and

8 PREPARATION OF SPECIMEN GRIDS

therefore difficult to handle and manufacture. Plastic sections which are rather unstable under the electron beam can be stabilized by a thin coating of carbon. The same stabilization effect is found when collodion (parlodion) films are coated with carbon.

The manufacture of carbon films requires a high vacuum evaporator and the actual process of "carboning" is discussed in the next chapter. Evaporation of carbon is done by heating rods of carbon to incandescence (to do this a current is run through two rods acting as $+$ and $-$ poles) in a vacuum. Carbon is radiated as small particles in all directions from the points of the two opposing rods. The amount of carbon radiated can be determined by watching the coating of a clean porcelain dish on which a drop of vacuum grease has been placed. The ungreased region will slowly blacken, while the white portion underneath the grease will not and the two regions can be compared.

A carbon film can be prepared by coating a plastic film. A plastic film is prepared as in Method I above, grids are placed on the film and the floating film is picked up so that the grids are now under the plastic sheet. Then, in the high vacuum evaporator, the film is coated with carbon until the desired thickness is achieved. Dissolve the plastic layer away by soaking the slide in absolute acetone for 2–3 min at 45° C (for parlodion), or chloroform for 2–3 min and then in ethylene dichloride for 2–3 hr (for formvar). The individual grids with their carbon films can be picked up and placed in a petri dish. Because of the delicate nature of carbon films one should keep all solvent dishes covered to reduce evaporation and avoid rough handling of the grids.

Carbon films can also be made by a "stripping" technique which requires freshly split mica sheets. Strips of mica, about 3–4 cm in width and 10 cm long are split so that a clean, freshly exposed surface results. These strips are placed in a high vacuum evaporator and coated with carbon as described above. The coated mica strip is then carefully dipped into fresh, clean distilled water (in a large finger bowl or pan) at an angle of about 15° to 25° to the water surface. Frosting the surface apparently helps to separate the carbon sheet from the mica strip. Grids are brought from below the floating carbon film up through it, which causes a film to be deposited on the grid. For this procedure, forceps holding the grid must be finely ground and only a small region of the grid edge held.

9 HIGH VACUUM EVAPORATION, REPLICATION, AND PARTICULATE SPECIMENS

The Evaporator 9.1

The high vacuum evaporator is used for preparation of biological material for electron microscopy in several techniques, specifically freeze etching, replication, particulate specimen examination, and coating-film manufacture. Basically, the evaporator is used to create a vacuum of about 10^{-6} torr (1 torr equals 1 mm of Hg). In this vacuum, metal shadowing with heavy metals and carbon coating of particulate carbon can occur. Most modern evaporators are controlled by a simple, single valve system or are automatic and all have a large bell jar where the vacuum is created and the coating equipment resides (Fig. 9-1).

Carbon Evaporation

Carbon coating is basically evaporation of fine particles of carbon onto a surface. A vacuum is needed to achieve this since oxidation and unequal spreading of the carbon particles will result when collision with gas particles occurs. The carbon rods are usually sharpened to permit a fine point-to-point contact (Fig. 9-4) and mounted in electrode holders to permit current to be passed through them (Fig. 9-2). As the site of the point-to-point contact becomes hot, the carbon particles fly off. The finer the carbon points, the finer the carbon particles that are produced and also fewer large carbon particles result.

Carbon coating is useful for strengthening collodion films, the manufacture of carbon films, and for strengthening replicas and plastic sections. Do not place the material to be coated directly under the rods as small pieces of carbon might fall off. The amount of carbon being deposited can be measured by placing a small piece of porcelain dish with a drop of vacuum grease on it in the path of the carbon. The white porcelain under the grease is then compared with the gradually blackening porcelain dish.

Metal Shadowing

Evaporation of heavy metals (palladium, platinum, and gold) from a point source at an oblique angle causes buildup of the

9 EVAPORATION AND REPLICATION

Figure 9-1 A photograph of a Kinney high vacuum evaporator. The investigator is holding a protective cage which fits around the glass bell jar. Inside the bell jar, where the vacuum is created, are the carbon evaporation and metal shadowing electrodes. The two large gauges shown on the right side of the evaporator are for reading the vacuum in the instrument (discharge and thermocouple gauges). The switches and rheostat on the left section of the front panel control the pumps and electrodes.

Figure 9-2 Carbon rod arrangement for carboning. The two rods, one with a fine point and the other with a blunt tip, are positioned under tension against one another. Current passing through the two rods causes the carbon to evaporate at the point. The carbon is deposited on the cleaned fragment of porcelain with a grease spot so that the amount of carbon evaporated can be determined.

THE EVAPORATOR 9.1

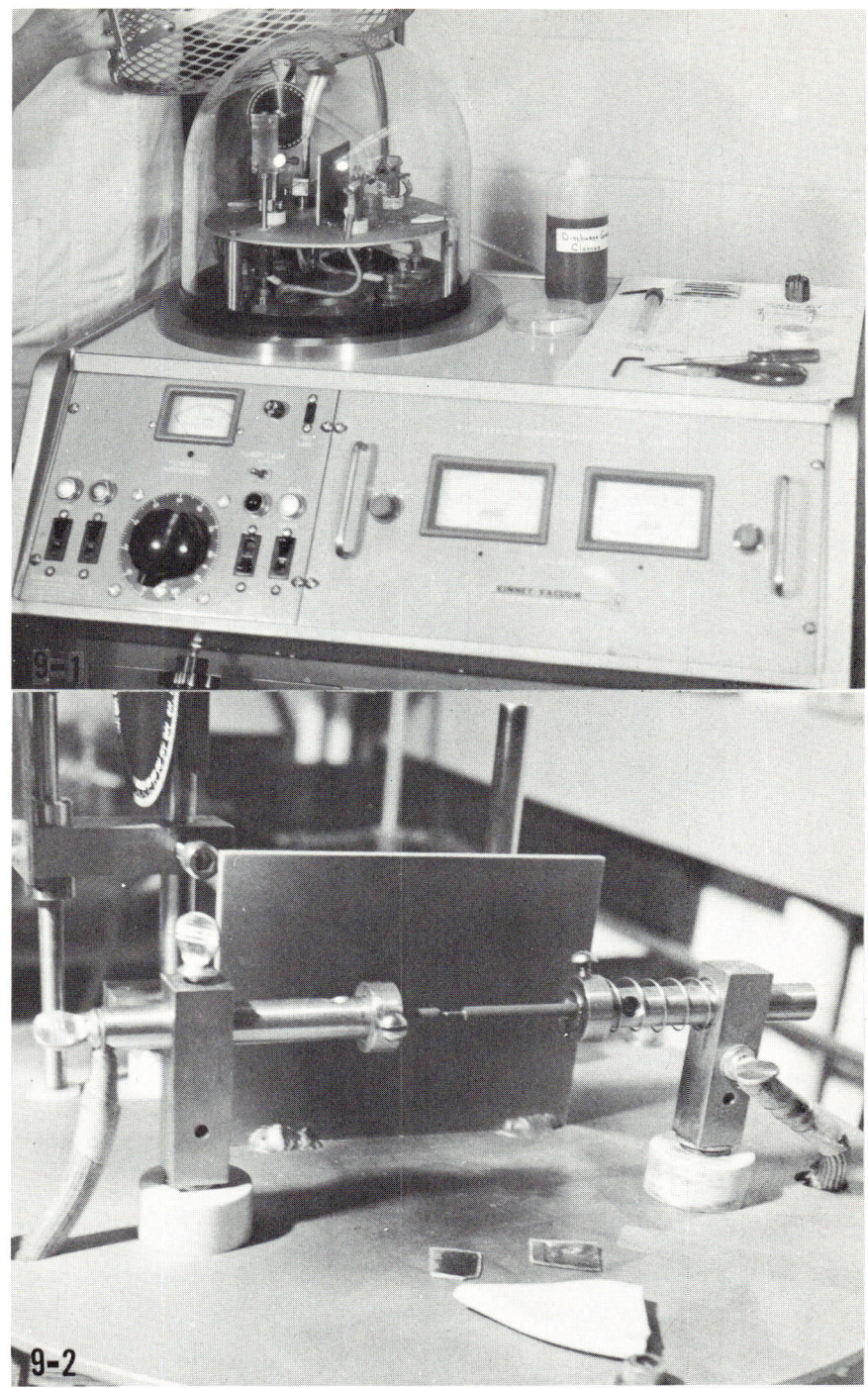

9 EVAPORATION AND REPLICATION

metal on the sides of the specimen contours which face the source (Fig. 9-3). This will create "drifting" or shadowing and cause a three-dimensional effect in particulate specimen images. The amount of metal deposited (that is, the thickness) can be determined by the following formula, and this in turn is used to determine heights of the contours in the specimen.

$$W \sin t = 4\pi d^2 p x$$

where W is the weight of metal; p the density of metal; d the distance between specimen and metal; t the angle of specimen to metal; and x the thickness of the metal layer on the surface.

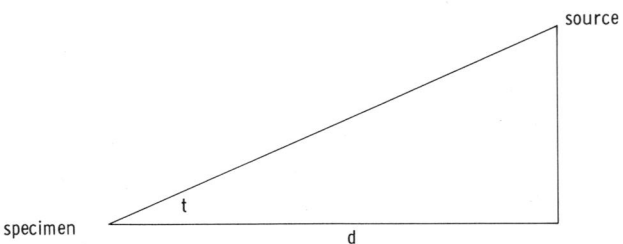

Shadowing is done whenever fine detail of whole mount material and freeze etch surfaces are desired. Of course, as with carbon coating, this must be done in vacuum to obtain sharp shadows. Usually the metal is wrapped around a tungsten wire (Fig. 9-4) or placed in a tungsten basket attached to electrodes (Fig. 9-3). Common metals used are chromium, gold, uranium, palladium/platinum (1:4), and platinum/carbon. Tungsten wires should first be cleaned with a NaOH solution and wiped dry in acetone to remove any oil coating (Fig. 9-4).

9.2 Replication

One of the first techniques employed for the preparation of material for electron microscopy was replication. Consequently, quite a bit of information and lore surrounds the manufacture of replicas (Bradley (1961)). Basically a replica is a copy of the specimen surface and consequently the potential resolution of the material to be copied is critical if a fair copy (or micrograph) is to be achieved. Replica methods are used in electron microscopy for the study of thick materials too opaque for resolution in the electron microscope, materials more conveniently sampled *in situ,* or materials that must not be altered or destroyed. Replication is also used in freeze etch techniques.

REPLICATION 9.2

Resolution of replicas is limited by the particle size of the replicating material (for example in plastics, the limitation is the molecular size of the monomer). Most materials give a resolution no better than about 10 nm. However, a thin metallic film evaporated on a surface to be studied can be resolved down to about 3 nm, and a carbon replica may resolve particles down to 2 nm (Table 9-1).

TABLE 9-1 Resolution of materials employed in replication.

Materials	Approximate resolution in nanometers
Parlodion	5–10
Formvar	5–10
Silica	3–5
Silicon monoxide	3–5
Evaporated metals	3–5
Evaporated carbon	1–2

If the specimen surface to be studied can be destroyed in the process, then collodion, formvar, or other plastics, or carbon may be used to coat the surface of the object, and the object is later dissolved from below the replica. If the surface of the object must be preserved, then plastics are generally used and stripped from the surface.

Replica techniques can be divided into three categories—one-phase, two-phase, and three-phase, depending on the number of copies to be made (Table 9-2). One-phase replicas are generally the most accurate, since two-phase and three-phase replicas lose resolution with each copy. In many instances, bits of the superficial layer of the original material become embedded in the replicating media and remain in the first copy. These pseudo-replicas offer advantages in that bits of actual material can be examined directly in the electron microscope for structural and diffraction patterns.

The variables involved in preparing replicas of any particular material include choice of method, choice of material for each replica (if more than one is used), the method of application or replicating substance, and the method of separation of the replica from the object. Because of the fragility of biological specimens, the most successful replicas and pseudo-replicas made from surfaces and sections are evaporated carbon and metal.

9 EVAPORATION AND REPLICATION

Figure 9-3 Metal shadowing device showing the coiled tungsten wire attached to the two electrodes. A small amount of palladium-platinum alloy is wrapped around the coiled portion of the tungsten wire and a current is passed through the wire. The alloy is evaporated off and will coat the porcelain fragment and any material (replica) at a pre-set angle (20° here). The dab of vacuum grease on the porcelain fragment can then be used to determine the amount of metal desired.

Figure 9-4 A display of materials and tools used in setting up for carboning or shadowing. From left to right at the far back are some presharpened carbon rods, a carbon rod sharpener, cleaned tungsten wire, a wood screw for coiling the tungsten wire, a coil of palladium-platinum wire, and common tools. A solution of discharge gauge cleaner should be available to permit frequent cleaning of the gauge.

REPLICATION 9.2

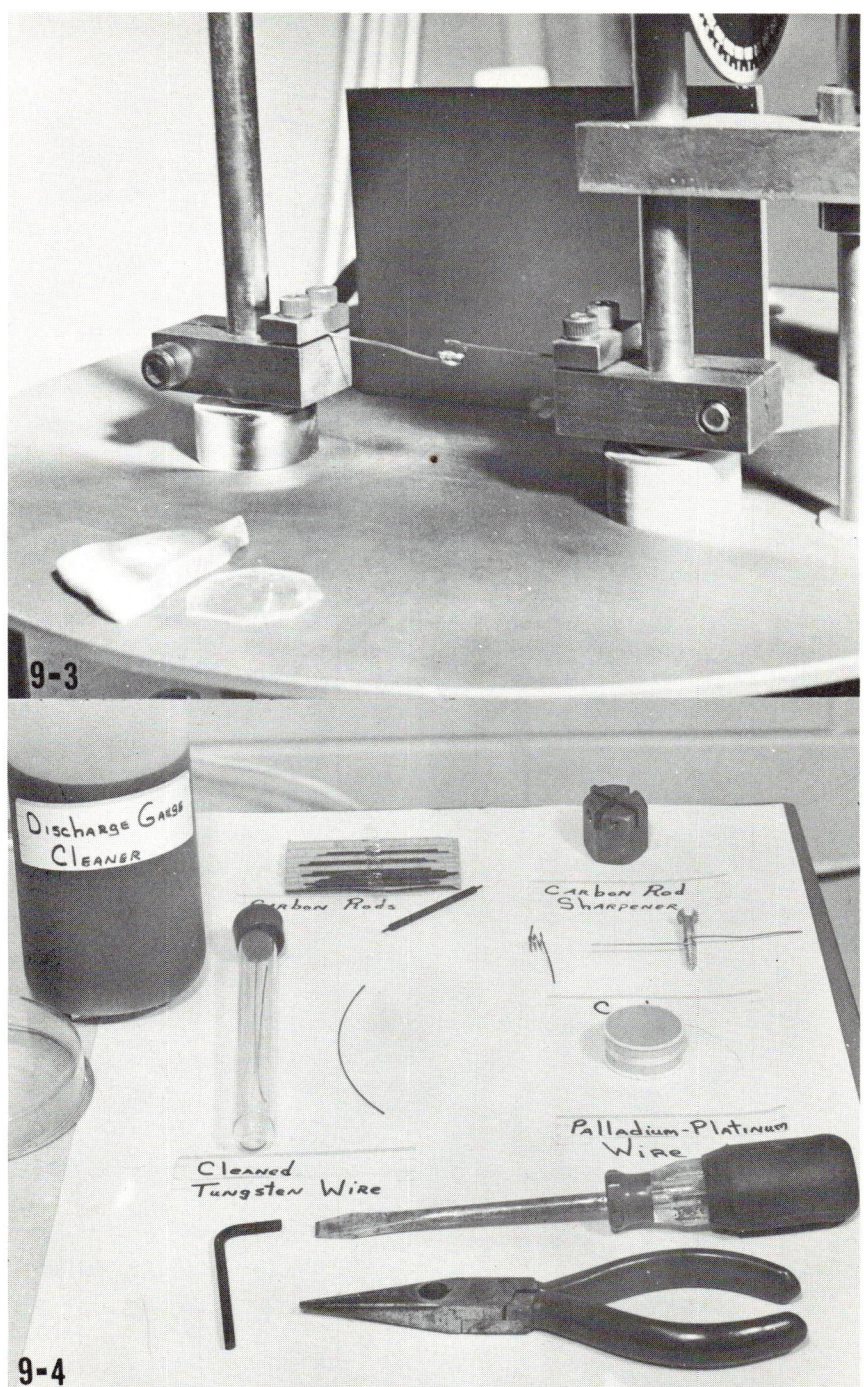

9 EVAPORATION AND REPLICATION

TABLE 9-2 Types of replicas.

Category	Type of replica produced	Materials used
One-phase	Negative copy	*Object:* Soft, friable, dry surfaces (e.g., plant cuticle). *Replica:* Evaporated silicon monoxide, carbon, and palladium-platinum.
Two-phase	Positive copy	*Object:* Solid, non-reactive materials, wet or dry (e.g., cell walls). *Replica:* Plastic coating for first phase, and evaporated carbon and metal for second phase.
Three-phase	Negative copy	*Object:* Highly porous surfaces (e.g., metal surfaces). *Replica:* Plastic block under high pressure; second replica, dilute plastic; third replica, evaporated metal from several point sources.

A number of replication techniques are described in the literature (Bradley (1961) and Wyckoff (1949)). However, only two, two-phase replica techniques are presented in detail here—the wet replica technique of Wardrop (1964) and the cellulose acetate replica technique of Henderson (1969). Both these techniques are simple to use and have yielded good results.

Henderson (1969), Cellulose-acetate replica technique The following materials are needed:

1. sheets of cellulose acetate, at least 0.5 mm thick and 25 by 25 mm square;
2. glass microscope slides;
3. acetone;
4. strips of transparent tape; and
5. 10% solution of polyvinyl alcohol that has been heated and stirred until the solution clears.

The squares of cellulose acetate are mounted on a microscope slide and the specimen to be replicated is embedded in the acetate by partially dissolving the acetate with acetone and pressing the specimen in. After the cellulose acetate hardens, the specimen may be trimmed with a razor blade or scalpel. The tissue is outlined with strips of transparent tape so that a shallow sink or well is made and this is filled with drops of 10% polyvinyl alcohol solution. After the polyvinyl alcohol is thoroughly dry, it is stripped off the specimen. This is the replica.

The polyvinyl alcohol one-phase replica is then shadowed with a heavy metal and carbon coated (in an evaporator). The plastic is dissolved away from the metal-carbon two-phase replica by floating the replica in a hot water bath at 97° C for 15 min. The two-phase replica is then picked up with specimen grids and examined.

The original embedded specimen can be reused for further strippings of polyvinyl alcohol. The carbon-metal replica can be examined under a dissecting microscope to determine the exact region of interest and the rest can be cut away. This permits picking up of only the interesting portion of the replica.

Wardrop (1964), Wet replica technique The following materials and preparations are required:

1. Two to 3 cm squares cut from sheets of celluloid are dried under gentle pressure in a 70° C oven for 3 days followed by 1 day in 100° C oven to remove camphor in the plastic (Fig. 9-4).
2. A solution of partially polymerized methyl methacrylate is made by taking 20 ml of of methyl methacrylate with inhibitor removed (see chapter 5, methacrylate plastics) and 0.2 ml of benzoyl peroxide and heating at about 70° C until the solution becomes viscous. Watch carefully and stir vigorously as the solution polymerizes quickly beyond this point (usually in about 20 to 30 min).
3. The partially polymerized methacrylate solution is then poured on the dried celluloid squares and allowed to set overnight in a 70° C oven.
4. A saturated solution of stripping plastic is made by heating the hydrolyzed powder of polyvinyl alcohol in a double boiler and stirring until the solution clears with some material remaining in suspension.

To make a replica, the celluloid squares with the polymerized methyl methacrylate are clamped against the material with a spring clamp and placed in boiling water for 15 to 20 min. To insure even pressure over the face of the plastic, squares of glass are placed on either side of the plant material, a celluloid square with squares of rubber sheet are then placed over this setup, and then the entire sandwich is clamped together with a large spring clamp (Fig. 9-5). After removal from the water, the tissue is stripped away and a drop of polyvinyl alcohol is allowed to dry on the square in a 70° C oven. When dry, the polyvinyl sheet is stripped off and the process is repeated until the impression is free of any specimen material. The impression is then shadowed with metal (palladium-platinum) at an angle of about 20° and

9 EVAPORATION AND REPLICATION

Figure 9-5   Materials needed for the wet replica technique of Wardrop (1964) are cut squares of hard rubber, squares of glass, squares of celluloid coated with methyl methacrylate, and a solution of polyvinyl alcohol. In the foreground, the layout for a replica "sandwich" is seen and also a completed sandwich held together with a metal paper clamp.

Figure 9-6   The technique based on dissolution of the plastic away from the metal and carbon replica for the wet replica technique of Wardrop (1964) is shown. A series of solutions in the ratios of 50:50 toluene:acetone, 100% acetone, 50:50 acetone:water, 30:70 acetone:water, and 100% water are required to separate the replica from the plastic impression and to permit pickup of the replica squares by specimen grids. The replica is first scraped clean of all excess coating (metal-carbon) so that only the actual replica is present. Then the replica is scored with a sharp razor blade into millimeter squares. This scored and scraped replica is then placed in the 50:50 toluene:acetone solution where the plastic is dissolved in about 3 min. By gentle shaking, the millimeter squares of the replica will float to the surface. These are then transferred through an increasing water series. Finally, the squares of replica are picked up by touching a grid to them in the distilled water dish. The transfer screen is made of fine copper mesh attached to a wire ring (soldered) and mounted on a wooden handle.

REPLICATION 9.2

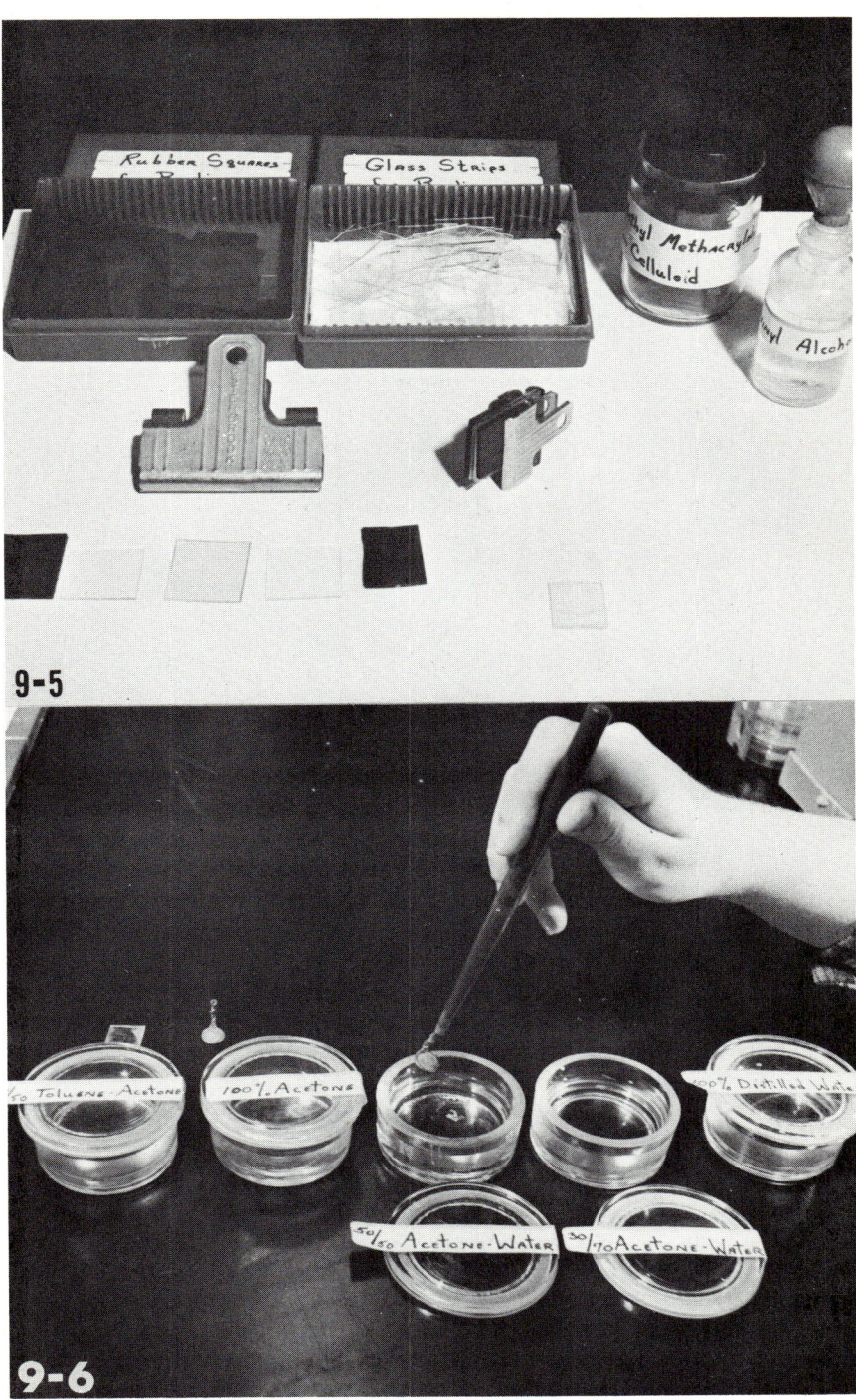

9-5

9-6

coated with a thin layer of carbon. The square is scraped with a razor blade to remove all metal/carbon that is not directly on the region of the replica. Then the replica is scored with a razor blade into 1 to 2 mm squares. The celluloid square with the coated replica is placed in a petri dish and dissolved in a 50/50 toluene/acetone solution. The small 1 to 2 mm squares of metal and carbon separate from the dissolving plastic and float to the surface as the dish is gently shaken. The floating squares of replica are transferred (Fig. 9–6) to 100% acetone, through 50/50 acetone and water, 30/70 acetone and water, and into distilled water. Transfer is done by use of a fine wire mesh screen attached to a wooden handle (Fig. 9-6). The replica squares are then picked up by touching a grid to the replica. Note that the squares will stretch out rapidly when transferred into increasing concentrations of water. To avoid breaking the squares, smaller concentration transfers to water can be made. Figure 9-8 is an example of the fine detail obtained through this technique.

9.3 Particulate Specimens

Prior to the techniques of ultrathin sectioning, the first procedures for specimen preparation for electron microscopy were limited to replication and studies of whole mounted material. The preparation of whole mounted material included fragmentation and mounting.

Fragmentation

In the case of whole mounted material, usually some sort of fragmentation is required to obtain small enough particles for viewing. Of course, fragmentation results in such a mixture of particles that the determination of which particles are being studied can be a real problem. Consequently, such studies were and still are limited to materials which represent a large proportion of tissue volume (mitochondria, chloroplasts, collagen fibrils, and cellulose microfibrils) and which can be fragmented without loss of their identity.

The process of fragmentation can be accomplished by either chemical disruption (such as dissolution of cementing substances such as pectic material of plant cell wall, middle lamella) or mechanical disruption (by ultrasonic vibrations of the chopping action of a blender, smashing of quick frozen material, or grinding between two places of ground glass). The material, if cellular in nature (e.g., microsomes, mitochondria, etc.) should be kept in a proper buffer with relation to tonicity and pH as well as temperature.

Plant cell wall fragments derived by either ultrasonic vibrations or by quick freezing of the cells and then by smashing the frozen material have been studied successfully in the author's laboratory. Small segments of plant material can be placed in an ultrasonic generator cup. Usually a cycle of about 1 megacycle per second for about 30 sec to 1 min will break up the tissue. Another technique, quickfreezing, produces larger more recognizable fragments of cell wall material. The segments of plant material were dipped in liquid nitrogen and then placed on a small, cooled (with dry ice) brass anvil and a spring-loaded brass block hovering over the anvil was struck down sharply. The resulting fragments are then gently washed off onto a small dish, and a drop of the fragments is placed on a coated grid.

Mounting

Particulate specimens are mounted on coated grids by a number of techniques. Usually, however, the material is in suspension in some liquid and can either be sprayed onto the grid (as with virus particles, bacteria, and fine cell fragments) or placed on the grid as a drop and allowed to dry down (as with cell wall fragments and bone fragments). Care must be taken to avoid a too rapid drying that will produce granular artifacts on the film surface of the grid. The concentration of the suspension must be watched. Typically, a solution is used of higher concentration than is necessary, so it is recommended that a number of dilutions from the original solution to a dilution of 1:1000 be tried.

The drop technique is best done with Pasteur pipettes. A single drop is placed on the coated grid and the drop allowed to dry down slowly in a slightly moist petri dish. The grids should be held or stuck down on a slide while placing the drop on them. Nebulization spray devices are available, or can be made, for applying a thin coat to coated grid surfaces. The latter technique works, of course, only with solutions of low viscosity and suspensions of small size particles.

In most cases of particulate specimen study, metal shadowing is performed to add contrast to the otherwise bland material. Figures 9-7 and 9-9 are examples of whole mount cell wall material. Examples of negative staining on particulate specimens can also be seen in Figures 10-8 and 10-9. It should be noted

9 EVAPORATION AND REPLICATION

Figures 9-7 through 9-9   Electron micrographs; all unit marks equal 1 μm.

Figure 9-7   An electron micrograph of a whole mount of a cell wall of *Dictyota flabellata,* a brown alga (Dawes *et al.* (1960, 1961)). The technique of metal shadowing the wall produces a three-dimensional aspect to the cellulose microfibrils and two pits seen in the wall. Folds and cytoplasmic debris stand out on the surface of the wall fragments. The cell walls were fragmented by ultrasonic maceration ($\times$ 5,000).

Figure 9-8   An electron micrograph of a replica (Wardrop's two-phase) of a developing conjugation tube (arrow) projecting from the filament of *Spirogyra,* a green alga (Dawes (1965)). The individual microfibrils on the surface of the filament wall and tube wall are easily seen, even at this low magnification. Resolution of this replica is about 5 nm ($\times$ 12,000).

Figure 9-9   An electron micrograph of a cell wall of *Apjohnia laetevirens,* a green alga, after sectioning and removal of the epoxy resin and subsequent shadowing with a heavy metal (Dawes (1969)). The alternating bands of microfibrils stand out because of the removal of the epoxy resin and subsequent shadowing ($\times$ 10,800).

PARTICULATE SPECIMENS 9.3

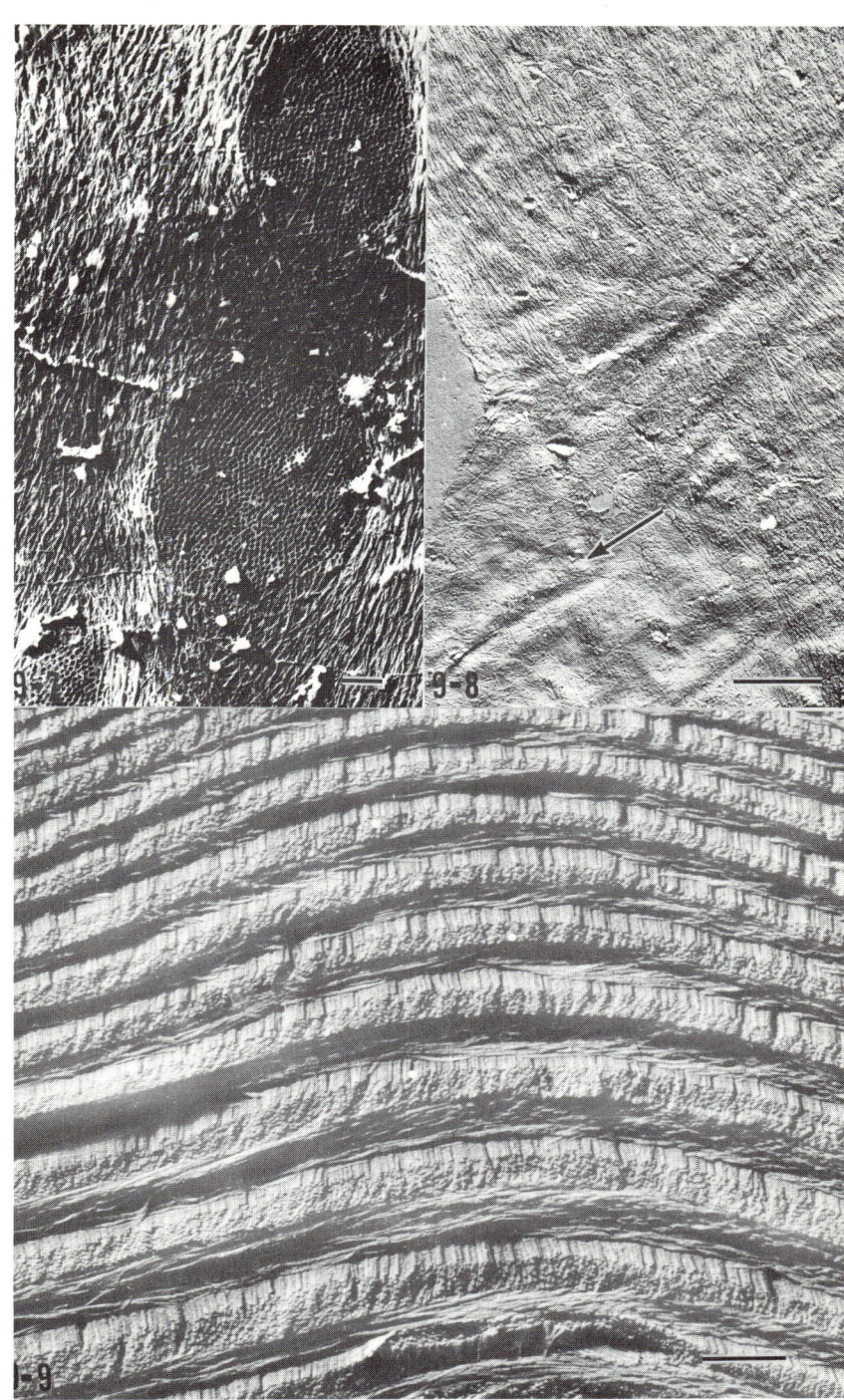

9 EVAPORATION AND REPLICATION

that any micrograph of shadowed material will really be positive, since the heavy metal prevents electron penetration and therefore the shadow appears as the material and not the shadow. It is traditional to make a contact negative (reversal) of the original negative, and this reversal is then used in printing.

10 STAINING

Introduction 10.1

Cells and tissues have a low intrinsic electron scattering power so that unstained sections yield little detail of their fine structure. To increase the scattering of electrons, which produces contrast in the specimen image, heavy metals such as lead, osmium, vanadium, and uranium are used to stain the sections. The thin sections are exposed to solutions containing salts or hydroxides of the heavy metal. The use of heavy metals for increasing electron scattering has been most important since the advent of epoxy plastics which lower the specimen contrast. Epoxies cause a large amount of electron scattering which increases the background noise in the image and thus reduces specimen contrast. This is not such a problem with methacrylate plastics which are more electron transparent.

The first stain used on sections was phosphotungstic acid, a molecule containing 24 tungsten atoms. This is now used primarily as a negative stain (Hall *et al.* (1945)). Alkaline lead hydroxide was introduced by Watson (1958) in a comparative study in which he used methacrylate embedments. Watson also introduced uranyl acetate in this study. These two are presently the most commonly used stains in electron microscopy. Sometimes the heavy metal solutions may complex with (react) or simply be absorbed by constituents of the cell. Uranyl acetate, for example, tends to combine with proteins and nucleic acids whereas lead hydroxide appears to be selectively absorbed by lipid materials.

Penetration of Stains 10.2

Methacrylate embedded tissue is the easiest to penetrate; usually maximum absorption takes place within 5 min. Penetration is most difficult in epoxies; araldite is the least penetrable, while maraglas is the most penetrable.

It should be pointed out that some staining can be done in the entire block during or after fixation and before dehydration.

10 STAINING

Uranyl acetate can be used (a 0.5% solution in the appropriate buffer) after aldehyde fixation or osmium postfixation. If the tissue is allowed to soak overnight in the cold (4° C), membrane structure will be enhanced and DNA fibrils in the nucleus and cell organelles will be preserved. Ruthenium red has also been used in a similar fashion (a 0.1% solution in the appropriate buffer) to trace out intracellular spaces.

Generally, most staining is done after sectioning and the stain is taken up directly by the sections. In many cases, one can improve the penetration of the stain by one of the following steps.

1. Pretreatment of the sections with organic solvents. Place grids with sections in the vapor of a solvent such as ethylene dichloride or amyl acetate for about 30 min. Do not use coated grids since the plastic coating will dissolve.
2. Float grids with sections face down on a 20 to 30% acetone or alcohol solution for about 30 min. This reduces the hydrophobicity of the plastic and allows better penetration of the aqueous stains.
3. Float grids with sections face down on stains which are dissolved in dilute alcohol solutions. This will accomplish roughly what was described in step (2).
4. Heat the specimens on the staining drops to increase stainability. Uranyl acetate penetration is especially helped by this technique.

10.3 Staining Procedure

Regardless of which stain is employed, the procedure for staining is a very simple one. All staining can be done on sheets of beeswax (dental wax) or clean, wax-filled petri dishes with at least one region (usually the center) forming a depression to hold pellets of NaOH (sodium hydroxide) which will absorb CO_2 during lead staining. After the sections have been mounted on the specimen grids, they are allowed to dry thoroughly before staining.

A drop of stain is then placed on the clean wax surface. The drop will round up because of the hydrophobic nature of the wax, and the grid, sections down, is placed on the drop and the dish is covered (Fig. 10-1). The length of staining time varies according to the stain used. Upon completion, the grid, held with a fine pointed forceps (for lead hydroxide stains), is alternately rinsed two times each in distilled water and $0.02M$ NaOH from squeeze bottles. The $0.02M$ NaOH is omitted for other stains. Rinsing grids after staining in permanganates should be done by alternating distilled water and 0.5% citric acid. Final drying of the grid can be done by gently touching the edge of the grid to filter paper or blowing off the liquid with condensed freon (containers and nozzle available from Ladd and French, Inc., Appendix I).

Stain Formulae and Reactions 10.4

The three most common stains are various forms of lead hydroxide, uranyl acetate, and the permanganates. Vanadium salts, phosphotungstic acid, and bismuth are also used. Compare Figs. 10-2, 10-3, 10-4, 10-5, 10-6, and 10-7 to see the results of most of these stains. In most cases, the staining reaction is related to the type of fixative used (Marinozzi (1963)). For example, if the tissue has been fixed or postfixed with OsO_4, lead hydroxide stains will react with RNA, cytomembranes, and starch (Fig. 10-4). Uranyl acetate will react with RNA and DNA, and consequently is a good stain for nuclei (Fig. 10-3). If the tissue has been fixed only in aldehyde, lead hydroxide stains appear to be limited only to RNA, while uranyl acetate will still react with RNA and DNA. The lead hydroxide stains are thought to react specifically with osmium molecules found on cytomembranes and carbohydrates by chelation of hydroxyl groups. Apparently lead salts may react with acid substances such as nucleic acids. However, less specific bonding also occurs with other high molecular weight units such as phospholipids (cytomembranes). For the permanganate stains, the dominant reaction is with cytomembranes and glycogen or starch (Fig. 10-5).

Lead Stains

These stains can be grouped into the lead hydroxide stains which are chelated (lead citrate, lead acetate, and lead tartrate) and those stains which are not chelated.

The general effects of lead hydroxide and chelated lead hydroxide are best seen in highly alkaline solutions. Apparently the maximum amount of cations available is at pH 12. A very high pH causes leaching out of the material in sections, and at still higher pH values these complexes are converted to anionic forms. Reynolds (1963) discusses the mechanisms of staining with lead salts and gives the following equation to describe the common feature of all stains using lead:

$$Pb(OH)_2PbX_2 \rightleftharpoons Pb(OH)_2Pb^{++} + 2X^-$$

where X is the chelating agent, if present.

10 STAINING

Figure 10-1 A demonstration of staining equipment used to stain sections mounted on grids. The rinse bottles (0.02N NaOH for lead stains; distilled water and duster for all stains) are used by squirting a gentle stream of water on the grid. The standard staining procedure is to use a petri dish (divided petri dishes with parafilm in two of the three sections). The use of a multiple staining dish as described for the Converse technique is shown on the right. The dish is mounted in a hand-made Plexiglas holder at an angle of 25°. Grids are clamped in a tweezer and placed in a staining cavity and then the grid is covered with the stain to be used. The stain is removed at the end by squirting the cavity with distilled water.

Figures 10-2 through 10-7 A series of sections taken of a dinoflagellate *Prorocentrum gracile* (Davis (1969)) which was fixed in 1% osmium tetroxide and embedded in Mollenhauer's (1959) mixture #2 of epon araldite. Each micrograph is taken of the same cell but different stains were used. All unit marks equal to 1 μm.

Figure 10-2 A general view of an unstained section from a cell of *Prorocentrum*. The square units are cross-sections of trichocysts (T). The chromatin material is seen as large fibrous bodies in the nucleus (N). Discoid plastids (P) are present at the periphery of the cell ($\times$ 14,000).

FORMULAE AND REACTIONS 10.4

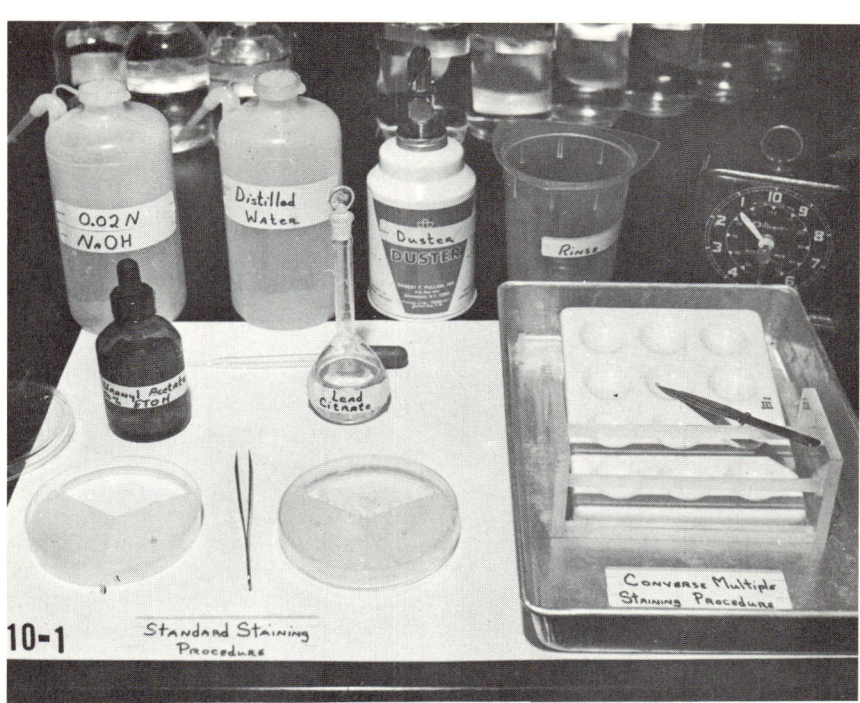

10-1 Standard Staining Procedure / Converse Multiple Staining Procedure

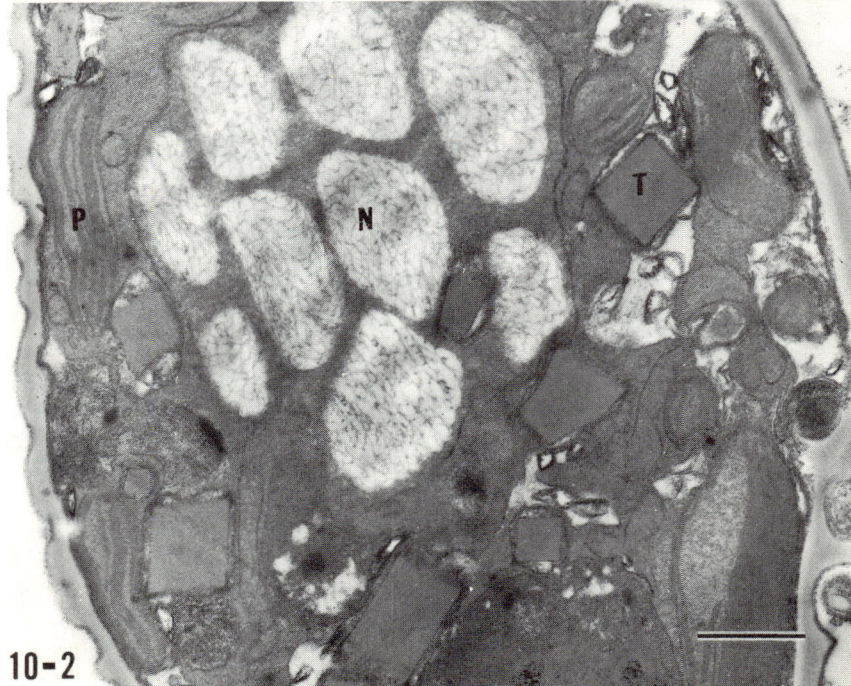

10-2

10 STAINING

Figure 10-3 A similar section of the cell *Prorocentrum* as in Fig. 10-2 stained with uranyl acetate for 30 min. The chromatin (Ch) material is much more obvious because of the stain ($\times$ 14,000).

Figure 10-4 Another section of *Prorocentrum* stained this time with lead hydroxide chelated with tartrate. The chelated lead stain has been picked up by various granules and the membranes of the Golgi bodies (G) seen around the nucleus ($\times$ 14,000).

Figure 10-5 This section of *Prorocentrum* was stained with 1% barium permanganate for 15 min. The cytomembranes and trichocysts (T) stand out because of the stain but little else appears affected ($\times$ 14,000).

Figure 10-6 An example of a section of *Prorocentrum* stained with vanadyl sulfate. The general appearance is that of a weak lead stain. The membranes show more contrast than that of an unstained section ($\times$ 14,000).

FORMULAE AND REACTIONS 10.4

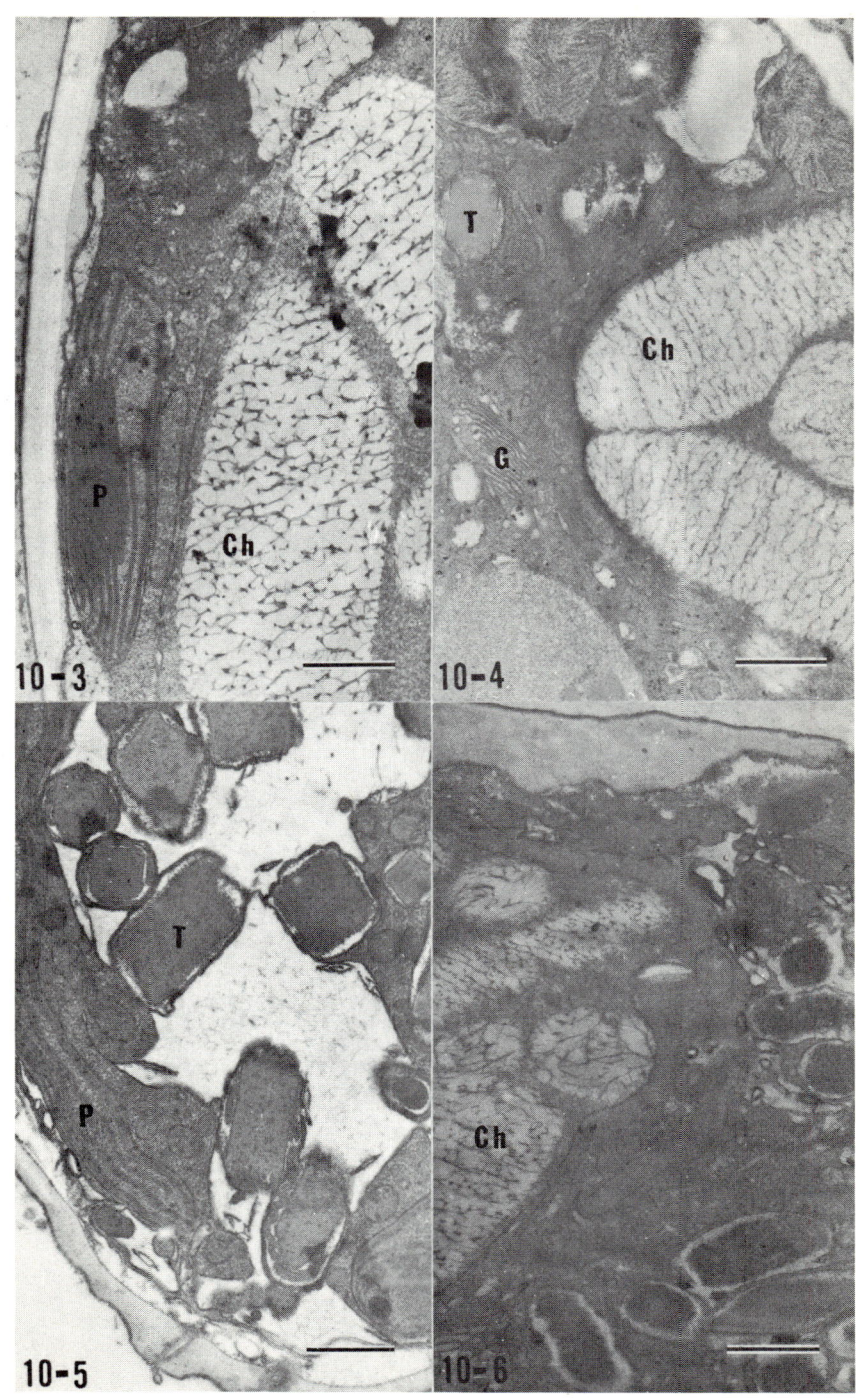

10   STAINING

Figure 10-7   An example of a double stained section. The section was first stained with uranyl acetate for 30 min, and then, after washing, it was stained with lead citrate for for 15 min. Here the combination appears to have improved the cell image much more than either stain would have alone ($\times$ 14,000).

Figure 10-8   An electron micrograph showing the effects of a negative stain, phosphotungstic acid in 0.4% sucrose on the phage F-1 of the bacterium *Clostridium sporogenes*. This stain forms a continuous background in which the phage stands out. Fine striations are seen in the phage tail ($\times$ 285,200). The unit mark equals 0.1 nm. (Micrograph by Dr. John Betz, University of South Florida.)

Figure 10-9   A similar preparation in which 2% uranyl acetate was used as the negative stain. Here, the stain also acted as a positive stain by reacting with the DNA of the phage head. Note the fine banding visible on the phage tail ($\times$ 219,400). The unit mark equals 0.1 nm. (Micrograph by Dr. John Betz, University of South Florida.)

FORMULAE AND REACTIONS 10.4

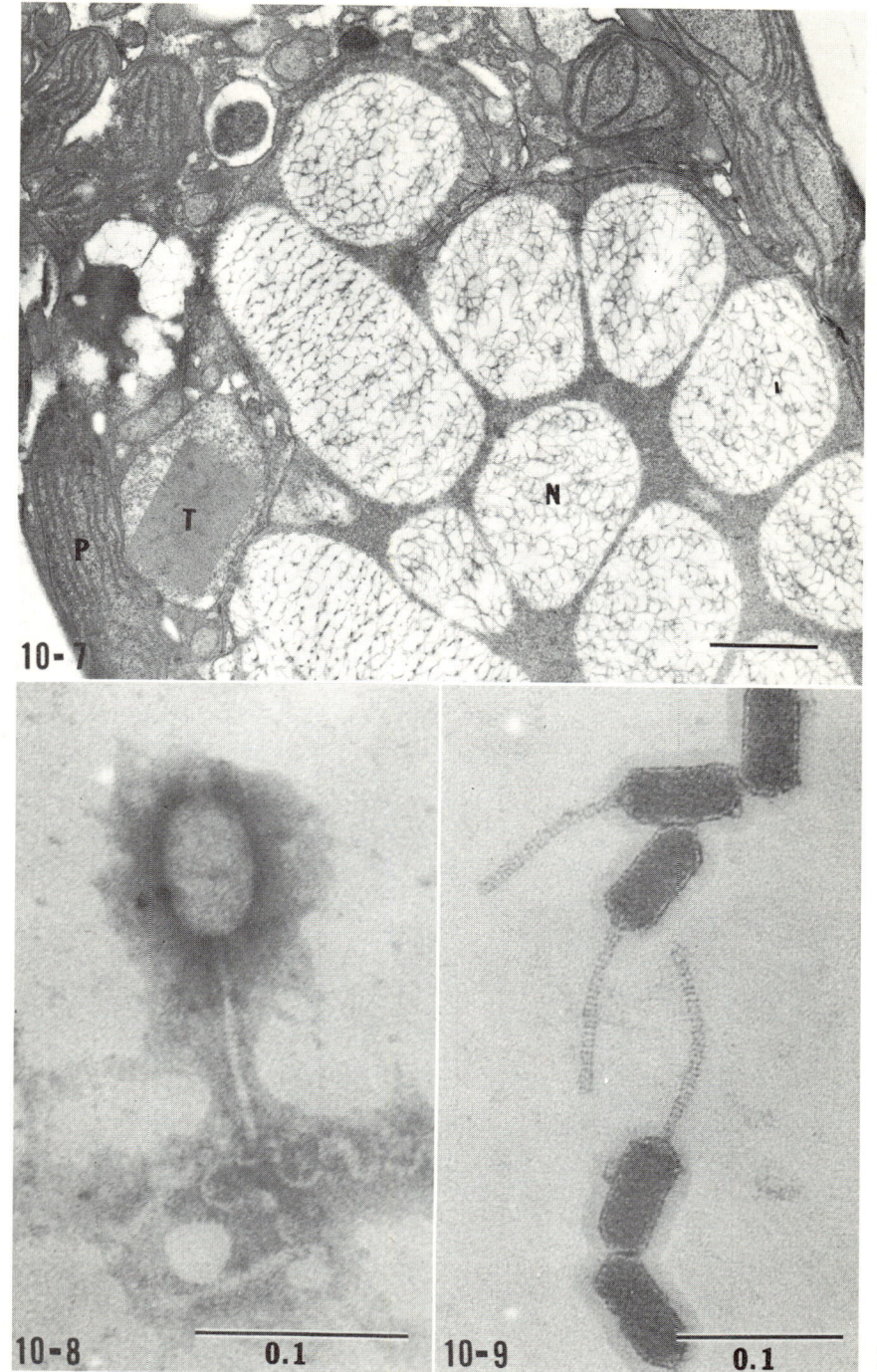

10-7

10-8 0.1

10-9 0.1

139

10 STAINING

The major problem with lead hydroxide is the formation of insoluble crystals of lead carbonate (a white powder) which contaminate the sections and ruin otherwise excellent results. This is the basic reason for using a chelating agent which binds up the lead hydroxide from carbon dioxide. Of course, such agents also reduce the staining ability of the lead yielding a stain with less contrast. Techniques for using unchelated lead hydroxide attempt to avoid carbon dioxide contamination and are described below for specific stains.

The following general procedures may be followed with all types of lead staining. All working surfaces should be cleaned before staining. Protect staining surfaces from the atmosphere by covering them and, if necessary, replace air with nitrogen. Submerge grids in the staining solution (lead carbonate floats) and then squirt away the staining solution before removing the grid. This technique of infinite dilution should prevent carbonate from settling on the grid. Use only freshly boiled distilled water when making up lead stains and rinse water. Use sodium hydroxide pellets (NaOH) in the closed petri dish to absorb CO_2. Keep the staining solutions in stoppered bottles or with a layer of oil on top to prevent carbonate formation, or in sealed centrifuge tubes (this permits centrifugation when a layer of carbonate accumulates on the surface). Caution: All **lead stains are dangerous if inhaled,** therefore **never leave glassware around which has contained these salts.** Wash everything after use, including the counter. Wash hands before leaving the room. Do not smoke or eat while staining or cleaning.

(1) Reynolds (1963), Lead hydroxide chelated with citrate

Preparation: In a 50 cc volumetric flask place:

lead nitrate ($Pb(NO_3)_2$)	1.33 g
sodium citrate	1.76 g
distilled water	30 ml

Shake vigorously for 1 min and then frequently for the next 30 min. Add 8.0 ml of 1N NaOH, at which time the solution becomes clear. Use only freshly made carbonate-free NaOH solution. Add distilled water to the 50 cc mark and mix by inversion. Lead citrate may be purchased in powdered form (Polysciences, Inc., Appendix I).

Staining: Grids are stained by floating them section side down on single drops of the stain for 5 to 30 min on a covered wax surface. After staining, the grids are rinsed in alternate streams of 0.02N NaOH and distilled water. Use longer times for epoxy resins, about 15 to 30 min. Results are usually free of carbonate, and leaching out is minimal despite the high pH of the stain. Glycogen aggregates stain intensely in thicker sections and appear as finely granular aggregates in thin sections.

(2) *Millonig (1961), Lead hydroxide, chelated with tartrate*
Tartrate as a chelating agent for lead hydroxide appears to be a stronger agent than citrate so that less carbonate is formed, but the stain is slightly weaker (Fig. 10-4).

Preparation: There are two methods depending on the availability of lead hydroxide.

Method 1:

NaOH	12.5 g
K-Na tartrate	5.0 g
add distilled water to make up to	50 ml

About 0.5 ml of this stock solution is diluted to 100 ml and heated with the addition of 1 g of lead hydroxide, and is then cooled and filtered. The final solution should be clear and have a pH of 12.3.

Method 2:

NaOH	20 g
K-Na tartrate	1.0 g
add distilled water to make up to	100 ml

About 1.0 ml of this is added to 5 ml of a 20% lead acetate solution, diluted 5 to 10 times, and then filtered to give a colorless solution.

Staining: The final solution from either Method 1 or 2 is kept in a stoppered bottle and is used as described for lead citrate stain.

(3) *Dalton and Zeigel (1958), Lead hydroxide chelated with subacetate* This stain uses subacetate as the chelating agent and appears to be somewhat selective for RNA. The pH is rather low, around 7.0, but appears to be more reasonable than the pH attained in lead acetate (see next formula, pH 5.9) or the pH of straight lead hydroxide (see below, pH 12).

Preparation: Add excess of monobasic lead acetate (Pb $(C_2H_3O_2)_2 \cdot$ Pb$(OH)_2H_2O$) to several milliliters of boiled distilled water (to remove CO_2) in a tiny container. Filter solution through fine filter paper in small funnel.

Staining: After the filtrate is clear, 3 to 4 drops are placed in a depression slide. Avoid breathing on the solution. Grids are

10 STAINING

immersed in the solution and a coverslip placed over the depression. After 5 min (for methacrylate) to 30 min (for epon) the grids are rapidly removed, drained, and rinsed by dipping several times in changes of distilled water.

(4) Dalton and Zeigel (1958), Lead hydroxide chelated with acetate In this formula, acetate is used as the chelating agent with a resulting pH of about 5.6. As noted, such a low pH value does not facilitate lead hydroxide formation, and consequently the stain is somewhat weaker than in other hydroxide solutions.

Preparation: In a tightly stoppered bottle, place an excess of lead acetate (so that undissolved crystals remain in the bottom) in boiled distilled water. Each time stain is removed, water is added to the glass stoppered bottle so that no air pocket remains at the top.

Staining: It is thought that the lead hydroxide does not form with this stain until it contacts the section due to the activity of acetate as a chelating agent. Grids are floated, section side down, on drops of stain on a clean wax layer in a petri dish. Stain for 5 to 15 min and watch for carbonate formation. Wash with squirt bottle vigorously 3 to 4 times using freshly boiled distilled water. Blow dry with freon. Wave grids, section side down, over vapors of 1 to 5% ammonium hydroxide; this is believed to convert the lead acetate to lead hydroxide. Such a technique also is thought to prevent carbonate formation.

(5) Karnovsky (1961), Lead hydroxide Karnovsky found that the stability of lead hydroxide as a stain was greatly improved if the Ph was kept around 12. Consequently, he recommends use of sodium hydroxide solutions in which lead hydroxide is made. These unchelated solutions are quite susceptible to carbonate formation and must be kept tightly stoppered.

Preparation: Two methods are given depending upon the availability of chemicals.

Method 1: Approximately 15 to 20 ml of 1N NaOH are placed in a flask with an excess of lead monoxide and boiled for 15 min. After cooling rapidly and filtering the insoluble material, the stock solution is kept in a closed container. To make the stain, dilute a given amount 50 to 200 times with boiled distilled water.

Method 2: An excess of lead monoxide is placed in 10–15 ml of a 10% sodium cacodylate solution and boiled for 15 min, cooled rapidly, and then filtered to yield the stock solution. The stain is made by dilution five times with 10% sodium cacodylate. Add 1N NaOH drop-wise with thorough stirring until clear (do not overtitrate).

Staining: The same procedure is followed as with lead citrate. Karnovsky believes that the effective staining unit is the plumbite ion and not the cationic radical as Reynolds stated. This appears to be more reasonable when one compares the results of effective staining between a chelated and an unchelated lead hydroxide solution.

(6) Watson (1958), Lead hydroxide

Preparation: Grind lead acetate (in a centrifuge tube grind 8.26 g lead acetate under 15 ml distilled water). Quickly pipette 3.2 ml 40% NaOH, and a dense precipitate will form. Sediment by centrifugation, discard the supernatant; rewash the precipitate with distilled water, centrifuge, decant, and resuspend sediment again. Recentrifuge and now retain the supernatant for a stain. This process can continue with the precipitate, two to three times more, adding the supernatant to the stain bottle each time. The preparation must be rapid since CO_2 reacts with the lead hydroxide, forming insoluble lead carbonate which is a common problem. Keep in a stoppered bottle. Pease (1964) recommends storage in a burette under oil to prevent $PbCO_3$ from forming.

Staining: Follow the same procedure as given in lead citrate with hydrophobic wax sheet or wax in a petri dish and individual drops of stain. Float grids, sections down, for 15 to 30 min in the stain in a closed container. This can be done in a CO_2-free atmosphere (nitrogen replacement).

(7) Converse (1969 *), Lead hydroxide

Preparation: Make a saturated solution of lead monoxide (PbO) in 1N NaOH (using carbonate-free NaOH and boiled water). Stir the solution while at a low boil and place in a stoppered bottle. This is the stock solution. Dilute the stock solution 1:50 with boiled distilled water and pour into centrifuge tubes and seal (with parafilm, etc.). Spin down any suspended matter and place tubes in a rack. These tubes contain the staining solution and will keep up to a month if properly sealed.

Staining: To stain use a multiple staining dish placed at a 20° angle in some sort of a rack. This entire unit should be placed in

---

* Mr. Bill Converse, Pathology Laboratory, University of Florida Medical School (personal communication).

10 STAINING

a shallow pan (Fig. 10-8). Pick up grid with a tweezers which can be locked, and place the grid and tweezers in one of the staining spots of the dish. Cover the grid with the staining solution. A locking type tweezers can be made by wrapping a piece of wire tightly around the handle. This wire ring can then be slid down the tweezers causing the tips to close. Use a syringe to remove the staining solution from the centrifuge tube. After completion of staining, rinse with boiled distilled water (infinite dilution technique) and blow dry with freon.

Uranyl Stains

(1) Watson (1958), Uranyl acetate As a general stain, uranyl acetate can be considered superior to phosphotungstic acid, but inferior to lead hydroxide. Proteins stain fairly well although staining of cytomembranes is not dramatic. The best reaction is with nucleic acids. In fact, uranyl acetate is commonly used as a fixative for nucleic acids, usually as a 0.5% solution in the same buffer as is used in the preceding aldehyde fixation.

Solubility is temperature dependent; at 15° C a 7.7% solution is saturated. A 5% solution in 50% ethanol is saturated at 15° C. Furthermore, the salt decomposes in hot water and contamination can build up in solutions kept around too long (2 to 3 months at room temperature).

Preparation: The 50% ethanol saturated with uranyl acetate is a good stain. To make the stain, an excess of uranyl acetate is placed in a small (5–20 cc) dark bottle containing 50% ethanol. Uranyl acetate dissolves slowly and a day should be allowed for the solution to reach saturation. As the solution is used, it can be replenished by adding more 50% ethanol to the bottle. However, each time ethanol is added, a day should be allowed for the solution to reach saturation before using the stain. Likewise, more uranyl acetate is added as needed. The stain is stored in the dark at room temperature, or in a dark bottle.

Staining: Filtering the stain just before use is optional. If filtering is not done, the stain can be removed from the bottle for use with a capillary pipette without disturbing the undissolved crystals at the bottom. The grids are floated section side down on the surface of the stain, or are immersed in the stain, section side up, for 15 to 30 min. After staining, the grids are picked up with a forceps, drained of excess stain on filter paper, and rinsed by dipping several times in two changes of distilled water.

(2) Uranyl nitrate This stain has been used with only moderate success by paralleling the procedure outlined by Reynolds for uranyl acetate; that is, dissolving 5% uranyl nitrate in 50% ethanol. The solution is a pale yellow and there is no excess. The results in staining compare with those of a weak solution of uranyl acetate. The stain itself forms small particles which do not obliterate the membrane systems in high resolution electron microscopy.

(3) Uranyl acetate-lead citrate double stain Using lead citrate after uranyl acetate gives an added effect, as though the uranyl acetate were acting as a mordant for the lead stain. Follow the procedures for uranyl acetate and lead citrate. This double staining procedure is the recommended standard procedure for a beginning student (Fig. 10-7).

Phosphotungstic Acid Stains

The phosphotungstic compound was first introduced by Hall *et al.* (1945) as a stain. This stain consists of 24 atoms of tungsten and is now primarily used as a negative stain in electron microscopy. Apparently, phosphotungstic acid binds with protein, although it does not attach to membranes well. Generally, it gives a "negative image" since membranes are light, while the background is dark.

Preparation: Prepare about 1% solution (distilled water or 40–50% alcohol), and stain for 15 to 30 min. After staining, dip grids in dilute water/alcohol, whichever was used in staining. Avoid scums by using very clean containers.

Permanganate Stains

Lawn (1960) suggested the use of potassium permanganate in a 1% solution, unfiltered and fresh. Either barium or potassium permanganate can be used although barium seems to yield a coarser image in cytomembrane staining than potassium. Both are strong oxidants so the duration of staining should last from 3 to 10 min. Rinse the grid in 5% citric acid, followed by distilled water, and dry. Avoid a surface scum resulting from oxidation of the stain by pipetting out from below.

Dirt may be found on the sections if stained as described. This is apparently caused by a reaction between the grid and the stain. Furthermore, sections are apparently "sensitized" to the electron beam, and, in some cases, are lost in staining, a feature of the permanganates. Best results are with coated grids.

10 STAINING

Parsons (1961) recommends using $KMnO_4$ (1%) as a block, stain for 10 to 15 min after dehydration in 100% acetone. Following this, the block is placed in a methyl acrylate-acetone solution (two drops methyl acrylate in 25 ml of acetone). This solution reduces and removes excess permanganate.

Vanadyl Salt Stains

The use of vanadyl salts was first demonstrated by Callahan and Horner (1964) using two different salts. These salts are considered to be general purpose stains which work well at low pH values. The results shown in Fig. 10-6 indicate that they are not as strong as the lead stains which they replace.

(1) *Vanadium sulfate*

Preparation: A 1% vanadyl sulfate solution (hydrated) is used at pH 3.6. A precipitate will form if the pH changes and green crystals of vanadyl oxide will form after about 2 weeks. Staining is for about 30 min, although longer periods do not contaminate the sections.

(2) *Vanadotomolybdate*

Preparation: Mix 20 ml of vanadyl sulfate with 80 ml of 1% ammonium heptamolybdate (molybdic acid). The resulting purple solution will become oxidized to a clear yellow liquid of vanadyl-molybdate (pH 3.2). Staining should be for 15 to 30 min and should be followed by a brief rinse with distilled water.

Ruthenium Red

Apparently this stain for light microscopy (pectic material) can be used during or after fixation in a 0.1% solution. The electron dense substance penetrates intracellular spaces and the resulting image shows dense regions of penetration between the cells.

10.5 Negative Staining

The use of particulate specimens (whole mounts) requires some sort of background to delimit the particles. The particles may be shadowed with metals or an electron dense substance may be allowed to dry down around the specimen. The term "negative staining" was coined to describe techniques in which an electron dense material, phosphotungstic acid or uranyl acetate, is deposited around particles using a wetting agent (bovine serum, starch, or sucrose).

NEGATIVE STAINING 10.5

A suspension of electron-transparent material such as micelles of cellulose, polysomes, bacteria, or virus, are mixed with a dissolved electron opaque substance such as uranyl acetate or phosphotungstate. A thin layer of the mixture is then dried onto a grid (coated with plastic) and the object becomes surrounded in a structureless electron opaque matrix. In the resulting micrographs the object then appears white, outlined by a dark background (Figs. 10-8 and 10-9).

Bovine serum base, phosphotungstic acid stain

Preparation:

 5N KOH:
KOH	28 g
distilled water to make	100 ml

 2% PTA:
phosphotungstic acid	2 g
distilled water	100 ml

The 2% phosphotungstic acid is adjusted to the desired pH (usually 4.5–7.2) with 5N KOH. The solution is then filtered and can be kept for a month or more, after which time it may develop some microscopic crystals.

 Buffered phosphotungstate (2%) with albumin (PTA with BSA):
buffered PTA	2 part
0.1% bovine serum albumin	1 part

This solution should be kept in the refrigerator to retard growth of fungi which might be observed in the final preparation when the object is being examined.

Staining: Take appropriate precautions with infectious microorganisms. One drop of a suspension of the specimen is mixed in a depression slide with two to four drops of PTA and BSA. A drop of the mixture is picked up by touching the support film side of the grid to the surface of the mixture. After 30 sec, the excess fluid is drained from the grid by touching its edge to filter paper, leaving a very thin film of liquid. It is allowed to dry, and the preparation is ready for examination.

The best pH value for the suspension medium and for the PTA is a matter of trial and error. The final choice is a compromise between the pH that seems physiologic for the specimen, best demonstrates the surface structure, gives some positive staining of internal structures, and gives the best spreading properties.

Starch base, phosphotungstic acid stain

Bovine serum is sometimes difficult to store and is much more variable to use than starch as a base.

Preparation: To 50 ml of distilled water add 1 g of phosphotungstate (sodium or potassium salt) and 2 to 3 drops of a starch solution. Add this staining solution to whatever particulate material is to be studied. Simply place a drop on a coated grid and dry. The starch solution is made by dissolving 1 teaspoon of normal potato starch in 500 ml of water and heating until clear. Based on the concentration of the particulate matter under study the final staining-particulate solution may be diluted.

Sucrose base, phosphotungstic acid stain This stain is made in a similar fashion to the previous formula. Instead of starch a 0.4% sucrose base is used (Fig. 10-8).

Uranyl acetate stain A 2% aqueous solution of uranyl acetate can be used directly as a negative stain by touching a grid coated with a suspension of particulate matter to a drop of the stain. The solution is then allowed to dry (Fig. 10-9).

10.6 Thick Section Staining

During ultrathin sectioning for electron microscopy, thick sections can be obtained for observation under the light microscope. Such sections are useful in determining the region of tissue being sectioned (for orientation) as well as for parallel study with the light microscope. Because of the opaqueness of the epoxy plastic it is recommended that the sections be stained, and because of the penetration problem with epoxy plastics certain stains are recommended.

Thick sections are cut on the ultramicrotome using the bypass technique, or whatever is recommended by the manufacturer. The thick sections are picked up from the trough with a flattened and shaved end of a toothpick and transferred to a small drop of distilled water on a microscope slide. The slide is gently heated until the drop has evaporated and the section is adherent to the slide. The heating also tends to expand the section, removing any folds.

Four stains are given here as recommended stains for epoxy embedded tissue.

Toluidine blue

Preparation:

Stock solutions:
    1% toluidine blue   (acqueous)
    2.5% sodium bicarbonate (Na_2CO_3)   (acqueous)

THICK SECTION STAINING 10.6

Make fresh stain each time by mixing:
  stock solution 1% toluidine blue   1 ml
  stock solution 2.5% Na_2CO_3 20 ml
and filter before use.

Staining: A drop of freshly made stain is placed over the sections on the slide. Allow to stain 10 to 20 min on a slide warmer at 47° C, using a microscope to check the degree of staining. Be cautious when using warmer temperatures which will decrease the staining time, because excessive heat will loosen the sections.

Rinse the stain off with distilled water. A slight destaining can be obtained with ethanol if desired. A small amount of acid in the alcohol will increase the destaining, but may loosen the sections. Dry and mount a coverslip using oil or a permanent mounting medium.

Methylene blue-azure II

Preparation:

Stock solutions:
  1% methylene blue in 1% borax
    methylene blue   1.0 g
    borax (sodium borate)   1.0 g
    distilled water   100 ml
  1% azure II
    azure II   1.0 g
    distilled water   100 ml

Mix equal parts of:
  1% methylene blue in 1% borax   10 ml
  1% azure II   10 ml

Staining: Add a pool of stain mixture and allow to stand 5 to 15 min. Warming the slide shortens the time, but occasionally this will loosen the sections. Rinse off excess stain gently with distilled water or 70% ethanol and allow to dry in air. Mount in immersion oil or permanent mounting media.

Basic fuchsin

Preparation:
  1% basic fuchsin in 50% ethanol
    50% ethanol   100 ml
    basic fuchsin   1 g

Staining: Add a pool of stain on the sections and allow to stand 5 to 15 min. Drain off excess stain and rinse with 70% ethanol

10 STAINING

(or distilled water). Dry in 40° to 60° C oven or in air. Mount in immersion oil or a permanent mounting media.

Crystal violet

Preparation:
    1% crystal violet
crystal violet	1 g
distilled water	100 ml

Staining: Add a drop of staining solution to the dried-down sections and let stand 1 to 10 min. Rinse thoroughly with distilled water and dry. Mount in immersion oil with coverslip.

11 PHOTOGRAPHY

Introduction 11.1

Photography is a most essential technique in electron microscopy, since the evidence obtained in ultrastructural studies is recorded on photographic materials. The image seen on the viewing screen of the electron microscope is caused by the fluorescence of zinc sulphide crystals which are being hit by electrons. Due to the relatively large size of the zinc sulphide crystals, the image appears grainy and the resolution is poor. For any detailed analysis and any permanent record, a photograph must be made.

Chemistry of Photography 11.2

The photographic process can be considered as a two-step process—developing and printing. In developing, the main concern is the production of a negative. In printing, the negative image is taken and made into a positive image or print. In both processes, the chemicals used are a developer (a reducer containing sulfate compounds) and a fixer (sodium hyposulfite, called "hypo" or "acid fix").

The photographic material usually consists of a glass plate, plastic film, or paper, coated with a layer of silver bromide crystals dispersed in gelatin (the emulsion). When these crystals are hit by light photons or high energy electrons, the silver is reduced in small regions within the silver bromide crystals. This initial reduction of silver is called the latent image of a negative or print. The latent image is developed into the visible image of a negative or print by the developer, which transforms the entire crystal of silver bromide into reduced silver. The crystals not affected by light photons or electrons are removed from the emulsion by the fixing agent after development.

Several crystals which have been transformed into reduced silver will form groups of silver grains in the emulsion. The density of these grains will cause the blackening of the film or print. The size of the silver grain formed depends on the size of the original silver bromide crystal. Emulsions with large silver bromide crystals will show greater sensitivity to the electron beam since larger crystals are more likely to be hit by an electron causing dense regions in the emulsion.

11.3 Exposure

The blackening or the density of the photographic emulsion depends on the electron intensity and the exposure time. The exposure (E) can thus be defined as the product of intensity and time; E represents the total incident energy. The graph (Fig. 11-1) shows that within a certain range of exposures a linear relationship exists between photographic density and exposure. The straight part of the curve reflects the greatest variation in density with certain changes in exposure. The angle between the ordinate and the straight part of the curve is the slope or gamma value (γ). The slope varies with different emulsions and also with type of development. Obviously, different emulsions are characterized by their γ values. A large γ value indicates that a small change in exposure results in a large change of density, a characteristic of high contrast emulsions. At very high γ values, the useful range for which an optimum exposure relation holds is very narrow, which means that it is necessary to expose the plates more exactly than at low γ values.

11.4 Film Speed

Sensitivity or the speed of the photographic material is characterized by the minimum exposure necessary to obtain a photographic density within a useful range of contrast. In ordinary photography, photographic materials are used that have been sensitized to a high degree to specific wavelengths. This sensitization, however, does not affect the sensitivity of the emulsions for electrons. Therefore, it is possible to use less sensitive photographic materials (for example, regular lantern slide plates), which are characterized by a fine grain emulsion.

Photographic emulsions are characterized by the grain size, the basis for fine and coarse grain films. An increase in sensitivity is, in general, associated with an increase in grain. In electron microscopy, as with other types of photography, a compromise must be made between speed and graininess of the film.

Electron Recording (DQE) 11.5

Kodak Techbits No. 2 (1965) and Hamilton and Marchant (1967) describe the theory of electron exposure of film. Basically, each electron may pass through as many as 100 photographic grains, losing energy each time to develop (reduce) at least some of the silver bromide crystals. The number of actual grains reduced depends on the particular developer and the development time. If, however, only one grain is developed then the emulsion records all the information of that electron.

Owing to its discrete nature, there is noise associated with the electron beam. That is, many of the electrons striking the emulsion do not carry any information and cause a background fog or emulsion granularity through random reductions of silver bromide crystals. The information-bearing electrons can be considered the signal. The signal-to-noise ratio will determine the image quality, and this ratio is equal to the square of the total number of electrons.

This can be expressed as

$$\frac{S}{N} = n^2$$

where S is the number of information-bearing electrons; N the number of random, scattered electrons; and n the total number of electrons of the electron beam.

The detective quantum efficiency (DQE) of a given emulsion is a method of determining how accurate the photographic record is. This can be expressed as

$$\text{DQE} = \frac{(S/N)^2_{\text{record}}}{(S/N)^2_{\text{beam}}}$$

where S is the optical density (signal) and N is the granularity (noise) of the emulsion for the photographic record; and S is the number of information-bearing electrons (signal) and N is the number of randomly scattered electrons (noise) of the electron beam. A DQE of 1 would indicate a perfect recorder; a DQE of 0.24 would mean that the signal-to-noise ratio is reduced by one-half in the recording step.

Recall the plot of density vs. exposure (Fig. 11-1). Different slopes are obtained if different emulsions are used. Another way of saying this is that with any given emulsion different plots can be obtained if the development time is varied. The two types of plates shown in Fig. 11-2 have characteristic curves, and the

11 PHOTOGRAPHY

Figure 11-1 A graph relating photographic density and exposure of three hypothetical emulsions. Emulsion A demonstrates a long linear relationship between exposure and density. Such an emulsion will permit predictable photographic density with a given exposure. Emulsion B has a shorter linear relationship between exposure and density. Also, the value is lower, indicating that small variations in exposure will not influence the photographic density of the film. Emulsion C shows almost no linear relationship between exposure and density. Such an emulsion would be quite difficult to calibrate for proper exposure.

Figure 11-2 A graph comparing two types of plate films used in electron microscopy. The graph shows the effect of developing for maximum, intermediate, and minimum grain. If processing for minimum grain (maximum resolution of film) the Kodak Electron Image Plates would require a slightly lower exposure to achieve the same density as Kodak Projector Slide Plates. The dashed line represents the Kodak projector slide plates, while the solid line represents the Kodak electron image plates. (Reproduced with permission from a copyrighted Kodak publication, Kodak Techbits, No. 2, 1966.)

DQE 11.5

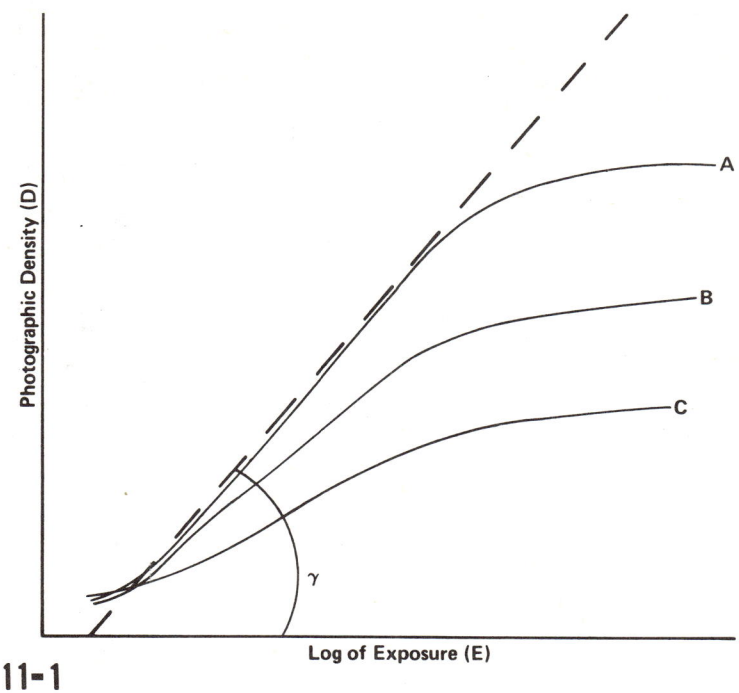

11-1

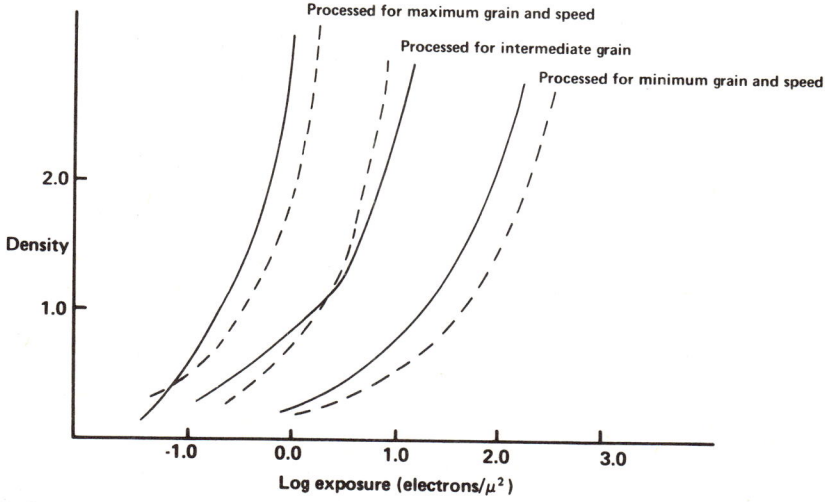

11-2

speed of the emulsion is related to the slope of the respective curve. To achieve the highest DQE possible, one should use the material with the lowest speed so that the largest possible number of electrons can be recorded (processed for minimum grain, Fig. 11-2). However, because of instability of the electron beam, the speed must be compromised toward a shorter exposure time.

If the electron beam were perfectly stable (no fluctuations or variations existing in the electron paths), then the only other limiting factor for a high DQE would be the "spread function" of the emulsion. This is the lateral electron displacement in the film which causes much of the noise or emulsion granularity. For electron microscopy, the spread function is not dependent upon the grain size of the emulsion (as in light photography), but rather on the gelatin content of the emulsion and the emulsion thickness. The higher the electron energy, the greater the spread function. In the 50 to 100 kV range of electron acceleration, the spread function is about 5 to 10 μm. The spread function will limit the possible enlargement because it causes emulsion granularity (as will the grain size of the emulsion). Because of the spread function of emulsions, longer exposures produce better images (more electrons are collected which record the image), and therefore slower films are preferred in electron microscopy.

A good rule of thumb is that as the enlargement of the microscope negative increases (when preparing prints), the grain of the micrograph should begin to become visible at the same time as the instrument resolution becomes a limitation (Kodak Techbits No. 2 (1965)).

11.6 Common Plates, Films, and Printing Papers

Projector Slide Plates

This type of glass plate backs a fine to medium grain emulsion and has been in use for a number of years. Since the emulsion is mounted on a glass plate, pumpdown time in the electron microscope is relatively fast. The emulsion itself is quite sensitive and development may be carried out in D-19 developer. The resulting image is of high resolution permitting enlargements of 10 to 15 times. This, of course, is because of the fine to medium grain size (Fig. 11-2).

Kodak Electron Image Plates

The major improvement of the electron image plates over projector slide plates appears to be the extremely fine grain emulsion with a slightly higher DQE when exposed to medium energy electrons (50 to 100 kV). Kodak Techbits No. 2 (1966) describes this emulsion as a thin layer of hardened gelatin mounted on glass plates so that susceptibility to abrasion is reduced. Unfortunately, this film is recommended for development in the rather unstable high resolution developer (HRP of Kodak) with an antifog supplement to obtain maximum resolution.

These plates offer only a slightly higher speed than projector slide plates with 100 kV accelerated electrons. The basic improvement is in the uniformity of the emulsion. Speed and granularity can be varied by changing the development time (Fig. 11-2). In other words, one can use high speed for unstable specimens and achieve a fairly fine grain capability (with high exposure) if the development is given maximum time.

The use of these plates permits a slightly higher DQE but also increases granularity which means the ultimate magnification of the plate will be limited (10 to 15 times enlargement usually). Consequently, if the development is carried out to achieve maximum contrast (and thus largest grain possible), then the DQE drops off slightly. The choice is between ultimate resolution with a slight loss in DQE or higher contrast with larger grain size.

Dupont Cronar Sheet Film

DuPont now offers an ortho/litho polyester base film (plastic thickness either 0.004 or 0.007 in.) to compete with the glass backed emulsions. Because of the lower handling costs and packaging, cut sheet film enjoys a cost reduction of three-quarters that of glass plates. Other advantages are that sheet film is not easily broken or destroyed, and the polyester base does not curl when drying or absorb water prolonging pump down time in the microscope. Experience with the film has shown that the development time can be lengthened to 6 min in full strength D-19 at 68° F. The resolution of the film is equal to that of other fine grain films, although the longer development may cause some grain.

Kodak Electron Microscope Film

The most recent sheet film produced for electron microscopy is a fine grain emulsion on an estar plastic base which does not require long pumpdown times in a vacuum and does not curl. The fine grain emulsion has a high DQE which means that the resolution of the emulsion is quite good (enlargement can also run to 15 times). The emulsion is blue-light sensitive so that a safelight in the Wratten Series OA range can be used. Furthermore, it can be developed in a standard developer such as D-19.

Kodak Fine Grain Positive 35 mm Film

This 35 mm film has been used for a number of years in electron microscopy because of its fine grain emulsion and very low cost (about $4.00 for a 100 ft roll). The emulsion is on a triacetate plastic base which does require a long pumpdown time in the vacuum. The emulsion is a slow type, about twice as long as projector slide plates. However, the 35 mm camera is mounted in the upper portion of the viewing chamber of the electron microscope so that the electron beam is about twice as intense. The film has a very fine grain emulsion which permits greater enlargement of the film record than that of projector or electron image plates; therefore, possible magnification differences between the films are slight. The DQE is slightly lower than that of projector plates but this is almost undetectable by the human eye (Kodak Techbits No. 2 (1965)).

Common Printing Papers and Processors

The printing papers in common use are the Kodabromide series (F1,2,3,4,5), the Agfa Brovira series (B1,2,3,4,5,6), the Kodak Polycontrast paper using polycontrast filters, and many other brands. Regardless of manufacturer, the highest contrast paper is the higher numbered one for the brand in question, while the polycontrast type paper is of one type and requires the highest numbered filter (#4). Choice of paper or filter to achieve the proper contrast level must be made with regard to the negative. Figures 11-3, 11-4, 11-5, 11-6, 11-7, and 11-8 are a series of properly exposed prints of the same negative using different contrast papers of the Kodabromide series (F1, 2, 3, 4, 5) and Agfa Brovira 6. Figure 11-7 is the correct contrast for this text.

Also available for printing are the stabilization processors and printing papers made by a number of manufacturers (Kodak, Ilford, Fotorite, and others). The printing papers contain a portion of the developing chemicals in the emulsion and both polycontrast type papers (requiring different filters) or graded contrast papers are available. After exposure, the print is simply run through a processing machine (costing from $200 to $800) into which the other developing chemicals are fed. For permanent prints, a final rinse in a fixing agent is required. The convenience of such a printing process (total time is less than 1 min) and the quality of the prints obtained indicates that the extra cost in equipment may be worthwhile.

Note: It is impossible to discuss the many other fine photographic materials available from other companies. A check should always be made with the local photographic dealer to determine other available supplies.

Darkroom Procedures 11.7

There are a number of general darkroom procedures that should be followed for printing or developing.

1. Keep the darkroom clean. Hyposulfite powder (fixer) or developer crystals will cause spots on the film and ruin it. Clean up after the work is completed, especially in areas where hands may have touched wall switches, doorknobs, and equipment. Wipe up spills on the floor and avoid splattering the chemicals.
2. Store all chemicals in the dark since most are light-sensitive. Keep the chemicals in clean, stoppered bottles labeled with the date and contents in clear, large print.
3. Do not expose boxes of printing paper to light, and date each box when opened. Record the date on the back side of the printing papers: exposure, paper type, and negative number.

Developing Negatives

Each type of film should be developed according to the recommendations of the manufacturer. A brief series of steps for the general procedure of developing Kodak 35 mm fine grain emulsion film is given below. If a film canister is used, practice loading with an exposed strip of film to insure correct loading. Do not touch the emulsion (usually the sticky side) since scratches may occur, especially when the emulsion is wet.

1. Develop Kodak Fine Grain Positive film in full strength Kodak D-19 at 20° C for 2 min. Film can be developed up to 7 min without too much fogging.
2. Wash the film for 15 sec in running water and fix for 5 min in hypo; save the used fixer, if used in a film canister.
3. Wash the film strip for 2 min in running water, then for 2 min in Hypo Clearing Agent, and 1 min in running water.
4. Place the film for 30 sec in a spot remover (Photo-Flo) or rinse thoroughly in distilled water and hang the strip up to dry.

11 PHOTOGRAPHY

Figures 11-3 through 11-8 An example of the difference between a series of contrast papers. An electron micrograph of the growing tip of a rhizome of a green alga, *Caulerpa prolifera,* was printed with proper exposure on a series of Kodabromide papers and Agfa Brovira B-6. Fixation of the green alga was with 6% glutaraldehyde followed by a postfixation of 0.6% potassium permanganate, both buffered in Sjöstrand's salt solution and veronal-acetate buffer. All micrographs are $\times$ 10,500, and the unit mark equals 1 μm. See Fig. 11-7 for key to symbols.

Figure 11-3 Printed on Kodabromide F-1. Note the lack of contrast and general blending of light and dark regions.

Figure 11-4 Printed on Kodabromide F-2. Contrast has improved some but there is very low tonal gradation.

Figure 11-5 Printed on Kodabromide F-3. Here the contrast of the vesicles is rather good although the cytoplasm/membrane contrast is still low.

Figure 11-6 Printed on Kodabromide F-4. Although quite similar to the previous print, the contrast has improved over the original.

DARKROOM PROCEDURES 11.7

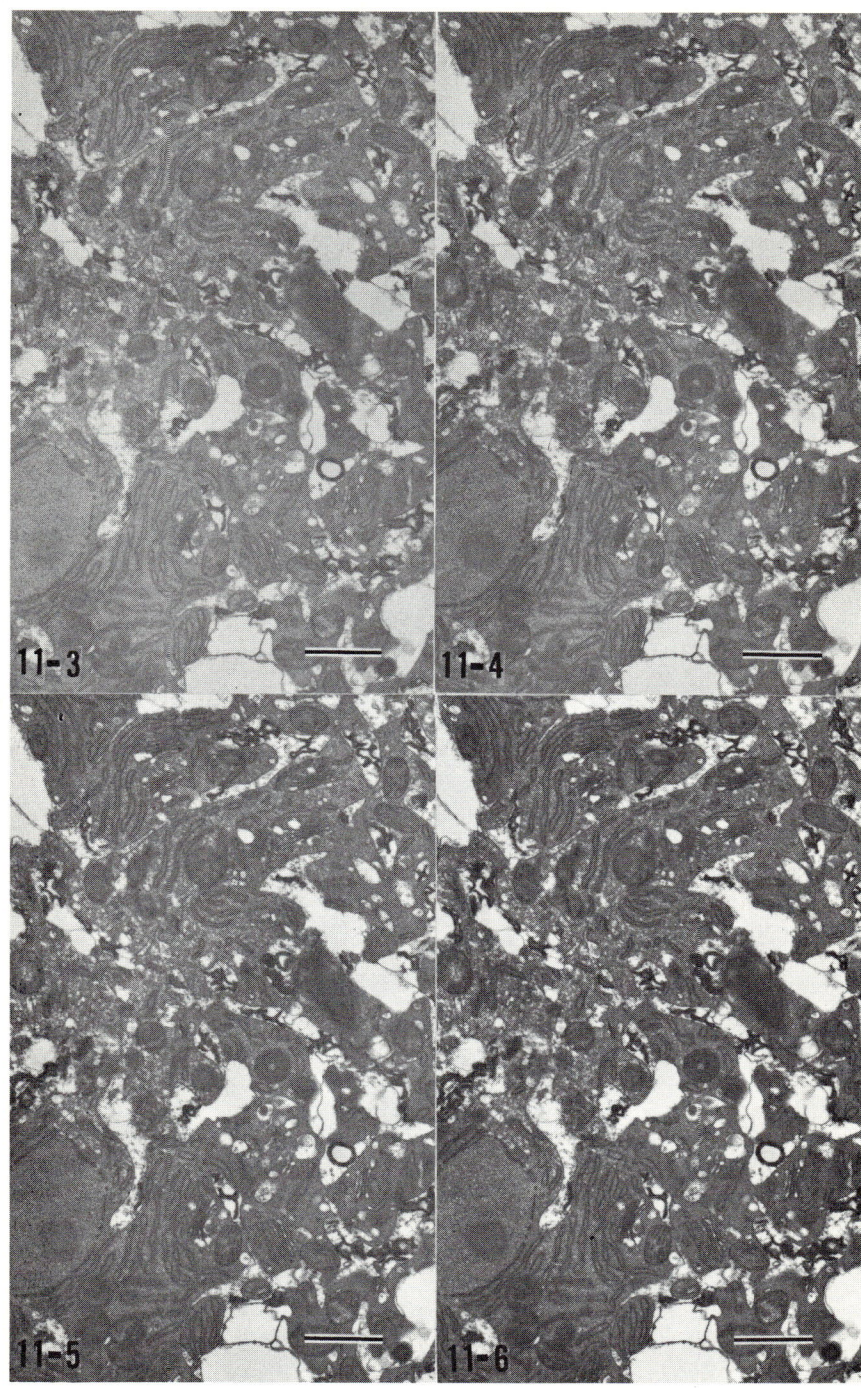

161

11 PHOTOGRAPHY

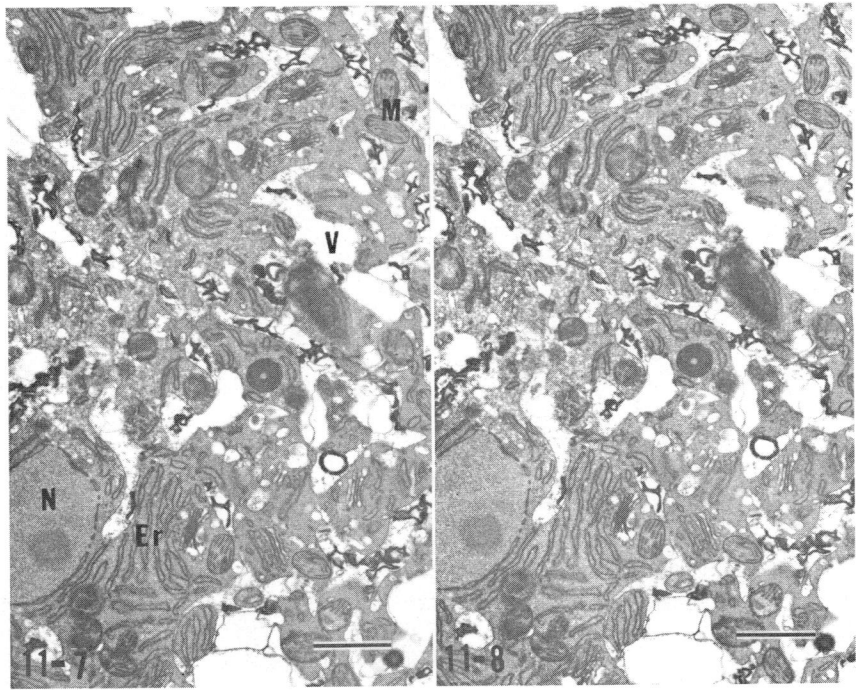

Figure 11-7 Printed on Kodabromide F-5. This print would reproduce well because of the high contrast of the micrograph, where N is the nucleus, Er the endoplasmic reticulum, M the mitochondria, and V the vacuole.

Figure 11-8 Printed on Agfa Brovira B-6. A print slightly too contrasty which makes the vesicles too white and causes some blurring (fogging) of the cytomembranes.

The film, when dry, should be cut up into sections and placed in envelopes or rolled up and stored in canisters.

Development of plates or sheet films (usually $3\frac{1}{4}$ by 4 in.) is done with either individual racks or group holders (up to 12 sheets or plates per holder). Take care not to scratch the emulsion, especially with glass backed plates. The following procedure is given for development of plates and sheet films.

1. Develop Lantern Slide Plates in full strength D-19 at 68° F for 3 min, Kodak Electron Microscope Film for 3 min, and Dupont Cronar Film for up to 6 min. Develop Kodak Electron Image Plates in HRP (High Resolution Processor) developer (diluted 1 to 4 with 5 g of Antifog added per liter) for 3 min at 68° F. A specific routine should be used in development, that is, agitation

DARKROOM PROCEDURES 11.7

and temperature should be constant to insure proper development. Agitate the development every 30 sec by picking up the plates or rack, tipping them to the right and then to the left and then replacing them in the developer. The entire agitation should not take more than 5 sec.
2. A red safe light is adequate (Wratten series I-OA).
3. Wash the plates for 15 sec (except for Electron Image Plates which require $1\frac{1}{2}$ min) and place in acid fixer (full strength) for 6 to 8 min.
4. Wash the plates for 2 min and then rinse for 2 min in Hypo Clearing Agent and then wash again for 5 min.
5. Finally, soak the plates or sheet film for 30 sec in a spot-preventing agent or simply rinse them in distilled water. If desired, the face of the plates can be wiped clean with a squeegee or soft sponge. Stand the plates or racks of film up to air dry.

Printing and Enlarging

This is the final step for obtaining micrographs from negatives produced in electron microscopy. Four basic solutions needed for printing are:

1. a dektol solution (1 part stock dektol and 2 parts water);
2. an acid stop bath (1% acetic acid);
3. a stock bath of hypo (fixer); and
4. a constantly running bath of rinsing water.

Constant checks should be made on the hypo bath. All solutions can be kept in normal developing trays with the water rinse in a sink.

The process of development can be divided into four steps. The first action is not visible to the human eye, for the paper in the developer remains white. In the second step, an image appears but lacks any tonal variation. Next, tonal gradation appears and, finally, there is a general fogging as the background density increases. Good prints are fully developed at the end of tonal gradation. The technician must develop a "good eye" since prints tend to exhibit more contrast in the developer under the safe light than actually is there. If a print is taken out too soon, the whites will be quite obvious. The print is considered a "pulled print" and is not usable (Fig. 11-11).

Light exposure is important, as seen in Figs. 11-9, 11-10, and 11-11, which show different exposures for the same contrast printing paper. The human eye is the best judge for proper exposure. The proper exposure for a given contrast paper or filter can be facilitated by use of a Kodak test density sheet or similar test. With a minute exposure on this sheet, the proper length of exposure can be determined.

Roughly, some image should be apparent in about 15 sec in a solution of 1 part dektol and 2 part water (step 2 of develop-

11 PHOTOGRAPHY

Figures 11-9 through 11-11 Three prints using Kodabromide F-4 paper demonstrating under- and overexposure as well as proper exposure of an electron micrograph of a cell wall of a red alga, *Ceramium* (Dawes *et al.* (1961)) ($\times$ 20,200).

Figure 11-9 A print which was underexposed so that the entire image appears rather dense and has a low contrast.

Figure 11-10 A print which was properly exposed. Note the fine detail of the cellulose microfibrils and the pit in the wall.

Figure 11-11 A print which was overexposed and had to be taken out of the developer earlier than desirable. The white areas are too sharp and thus fine detail is lost.

DARKROOM PROCEDURES 11.7

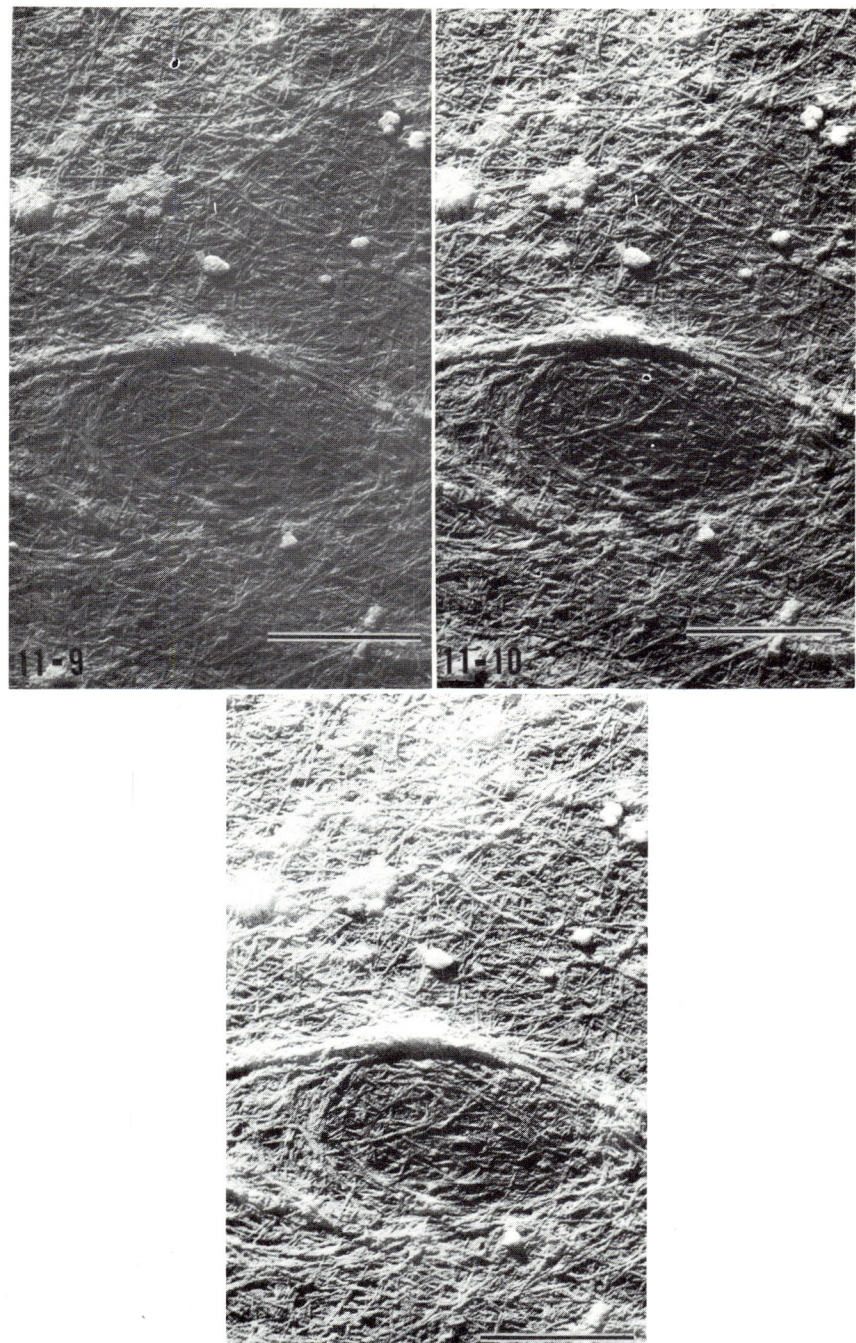

11 PHOTOGRAPHY

ment). A complete image with proper tonal distinction should be present within 1 to 2 min. In an enlarger with normal filament illuminations, the enlarging lens should be closed down about 2 to 4 f-stops back from wide open (brightest light). This f-stop is recommended because only the central and most distortion-free region of the lens is used and because the greatest depth of field occurs in the higher f-stops. Always focus the enlarger with the lens wide open where the brightest image is given and where the smallest depth of field occurs.

Dodging can be used to rid the print of unusually bright spots (low negative density regions) which would cause haloing or fogging of the adjoining image. These bright spots (perhaps caused by a hole in the plastic or a grid bar image) can be toned down by preventing that portion of the micrograph from being exposed as long as the remainder of the micrograph is exposed. This is accomplished by cutting out various forms to fit the bright region and then holding this form on the tip of a dissecting needle between the lens and the paper over the bright spot. During the exposure, the form or dodge is moved up and down until the shape just covers the bright region and then it is shaken to prevent an imaging of the form on the print and to cause a "toning in" of the dodged region.

The following procedure can be used when printing:

1. Identify the print by writing negative number, exposure date, and contrast number (paper or filter) on the back of the sheet before developing.
2. Determine the length and intensity of exposure by using a test density sheet. Place this test sheet over the selected printing paper and expose for 1 min. Check that the desired contrast level is obtained.
3. Develop the printing paper in Kodak dektol (1 part dektol to 2 parts water). Temperature should be around 20° to 22° C and length of development should be uniform, about 1 to 2 min.
4. Place the print in a stop bath (1% acetic acid) for 15 to 30 sec and then in the fixer for 6 to 10 min. Check the fixer for usefulness with a fixer check solution (Kodak Hypo Fix).
5. Wash the prints for about 1 hr, but do not leave the prints washing overnight or they may be damaged. If a clearing agent is used (Kodak Hypo Clearing Agent), the washing can be reduced to a few minutes.
6. Place the prints in a clean 1% glycerin solution (a print flattener or glossy solution can also be used) for 2 to 5 min, and then dry the prints in a dryer. Avoid overheating as this will cause shelling or cracking of the print gloss. Make certain the drying plates (usually ferro-type) are clean and spot-free.
7. After drying, place the prints under some weights to prevent curling.

REFERENCES

Baker, John R. 1965. "The Fine Structure Produced in Cells by Fixatives." *J. Royal Micr. Soc.* **48**:115–131.

———. 1965b. "The Fine Structure Produced in Cells by Primary Fixatives. 2. Potassium Dichromate." *Quart. J. Micr. Sci.* **106**:15–21.

———, and Luke, B. M. 1963. "The Fine Structure Produced in Cells by Primary Fixatives. 1. Mercuric Chloride." *Quart. J. Micr. Sci.* **104**:101–106.

Bourne, V. L., and K. C. Bourne. 1969. "A Simplified Technique for the Application of Nuclear Emulsion in Electron Microscopic Autoradiography." *Can. Jour. Bot.* **47**:1821–1822.

Bradley, D. E. 1961. "Replica and Shadowing Techniques," in D. Kay (Ed.), *Techniques for Electron Microscopy* (Charles C. Thomas, Springfield, Ill.).

Bullivant, S., and A. Ames. 1966. "A Simple Freeze-Fracture Replication Method for Electron Microscopy." *J. Cell Biol.* **29**:435–447.

Callahan, W. P., and J. A. Horner. 1964. "The Use of Vanadium as a Stain for Electron Microscopy." *J. Cell Biol.* **20**:350–356.

Cargille Laboratories, R. P. 1962. "Background and Instructions for the Use of the Cargille (NYSEM) Epoxy Embedding Kit." *Data Sheet EEK138.* (R. P. Cargille Laboratories, Liberty St., N.Y.), 10 pages.

Caro, G. C., R. P. van Tubergen, and J. A. Kolb, 1962. "High Resolution Autoradiography. I. Methods." *Jour. Cell Biol.* **15**:173–188.

Claude, A. 1948. "Studies on Cells: Morphology, Chemical Constitution, and Distribution of Biochemical Functions." *Harvey Society, N.Y.* **48**:121–164.

Cosslett, V. E. 1967. "The Future of the Electron Microscope." *J. Royal Micr. Soc.* **87**:53–76.

Dalton, A. J. 1955. "A Chrome-Osmium Fixative for Electron Microscopy." *Anat. Record* **121**:281.

——— and R. R. Zeigel. 1958. "A Simplified Method of Staining Thin Sections of Biological Material with Lead Hydroxide for Electron Microscopy." *J. Biophys. Biochem. Cytol.* **7**:409–410.

Davis, J. T. 1969. "Ultrastructural Studies of Several Marine Dinoflagellates." (M. A. thesis, Univ. South Fla., Tampa, Fla.).

Dawes, C. J. 1965. "An Ultrastructural Study of *Spirogyra*." *J. Phycol.* **1**:121–127.

———. 1969a. "A Study of the Ultrastructure of a Green Alga, *Apjohnia Laetevirens* Harvey with Emphasis on Cell Wall Structure." *Phycologia.* **8**:77–84.

REFERENCES

———. 1969b. *"Saprochaete saccharophilia:* Ultrastructure, X-ray Diffraction and Chitin Assay of Cell Walls as Aids in Evaluating Taxonomic Position." *Trans. Amer. Micr. Soc.* **88**:572–581.

——— and E. Rhamstine. 1967. "An Ultrastructural Study of the Green Algal Coenocyte *Caulerpa Prolifera.*" *J. Phycol.* **3**:117–126.

———, F. M. Scott, and E. Bowler. 1960. "A Light and Electron Microscope Study of the Cell Walls of a Brown Alga and a Red Alga." *Science* **132**:1163–1164.

———, ———, and ———. 1961. "A Light and Electron Microscope Study of Algal Cell Walls. I. Rhodophyta and Phaeophyta." *Amer. J. Bot.* **48**:925–932.

Echlin, P. 1968. "The Use of the Scanning-Reflection Electron Microscope in the Study of Plant and Microbial Material." *J. Roy. Micr. Soc.* **88**:407–418.

Fisch, W., and W. Hofmann. 1956. "The Curing Mechanism of Epoxy Resins." *J. Appl. Chem.* **6**:429.

Freeman, J. A., and B. O. Spurlock. 1962. "A New Epoxy Embedment for Electron Microscopy." *J. Cell Biol.* **13**:437–443.

Gibbs, S. P. 1962. "The Ultrastructure of the Chloroplasts of Algae." *J. Ultrast. Res.* **7**:418–435.

Glauert, A. M. 1961. "Section Staining, Cytology, Autoradiography and Immunochemistry for Biological Specimens," in D. Kay (Ed.), *Techniques for Electron Microscopy* (Blackwell Scientific Publications, Oxford, England), pp. 254–310.

———, and R. H. Glauert. 1958. "Araldite as an Embedding Medium for Electron Microscopy." *J. Biophys. Biochem. Cytol.* **4**:191–194.

Hall, C. E. 1966. *Introduction to Electron Microscopy* (McGraw-Hill Book Company, N.Y.), 397 pages.

———, M. A. Jakus, and E. O. Schmitt. 1945. "The Structure of Certain Muscle Fibrils as Revealed by the Use of Electron Stains." *J. Appl. Phys.* **16**:459–465.

Hama, K., and K. R. Porter. 1969. "An Application of High Voltage Electron Microscopy to the Study of Biological Materials." *J. de Micr.* **8**:149–158.

Hamilton, J. F., and J. Marchant. 1967. "Image Recording in Electron Microscopy." *J. Optical Soc.* **57**:232–239.

Heather, D. M., J. C. Hampten, and B. Rosario. 1961. "A Simple Method for Removing the Resin from Epoxy-Embedded Tissue." *J. Biophys. Biochem. Cytol.* **9**:909–910.

Henderson, W. J. 1969. "A Simple Replication Technique for the Study of Biological Tissues by Electron Microscopy." *J. Micr.* **89**:369–372.

Hillier, J., and M. E. Gettner. 1950. "Sectioning of Tissue for Electron Microscopy." *Science* **112**:520–523.

Holt, S. J., and R. M. Hicks, 1961a. "Use of Veronal Buffers in Formalin Fixatives." *Nature* **191**:832.

———. 1961b. "Studies on Formalin Fixatives for Electron Microscopy and Cytochemical Staining Purposes." *J. Biophys. Biochem. Cytol.* **11**:31–45.

Ito, S. 1962. "Light and Electron Microscope Study of Membranous Cytoplasmic Organelles," in R. Harris (Ed.), *The Interpretation of Ultrastructure* (Academic Press, N.Y.), pp. 129–149.

REFERENCES

Juniper, B. E., G. C. Cox, A. J. Gilchrist, and P. R. Williams. 1970. *Techniques for Plant Electron Microscopy* (Blackwell Scientific Publications, Oxford, England), 108 pages.

Karnovsky, M. J. 1961. "Simple Methods for Staining with Lead at High pH in Electron Microscopy." *J. Biophys. Biochem. Cytol.* **11:**729–732.

———. 1965. "A Formaldehyde-Glutaraldehyde Fixative of High Osmolarity for Use in Electron Microscopy." *J. Cell Biol.* **27:**137A.

Kay, D. (Ed). 1961. *Techniques for Electron Microscopy.* 2nd ed. (Blackwell Scientific Publications, Oxford, England), 560 pages.

Kellenberger, E., A. Ryter, and J. Sechaud. 1958. "Electron Microscopy Study of DNA-Containing Plasms II. Vegetative and Mature Phage DNA as Compared with Normal Bacterial Nucleoids in Different Physiological States." *J. Biophys. Biochem. Cytol.* **4:**671–678.

Kodak Techbits. 1965 No. 2. "Some Things Every Electron Microscopist Ought to Know" (Eastman Kodak Co., Rochester, N.Y.), pages 3–7.

Kodak Techbits. 1966 No. 2. "Electron Microscopist's Recompense" (Eastman Kodak Co., Rochester, N.Y.), pages 1–3.

Kodak Techbits. 1968 No. 3. "A Contrast Boost for Electron Microscopists" (Eastman Kodak Co., Rochester, N.Y.), pages 2–3.

Koehler, J. K., K. Muhlethaler, and A. Frey-Wyssling, 1963. "Electron Microscope Autoradiography. An Improved Technique for Producing Thin Films and Its Application to H^3-Thymidine-Labeled Maize Nuclei." *Jour. Cell Biol.* **16:**73–80.

Latta, H., and J. F. Hartmann. 1950. "Use of a Glass Edge in Thin Sectioning for Electron Microscopy." *Proc. Soc. Exptl. Biol. Med.* **74:**436–439.

Lawn, A. M. 1960. "The Use of Potassium Permanganate as an Electron Dense Stain for Sections of Tissue Embedded in Epoxy Resin." *J. Biophys. Biochem. Cytol.* **7:**197.

Luft, J. H. 1956. "Permanganate—A New Fixative for Electron Microscopy." *J. Biophys. Biochem. Cytol.* **2:**799.

———. 1959. "The Use of Acrolein as a Fixative for Light and Electron Microscopy." *Anat. Record* **133:**305.

———. 1961. "Improvements in Epoxy Resin Embedding Methods." *J. Biophys. Biochem. Cytol.* **9:**409–414.

Maaløe, O., and A. Birch-Anderson. 1956. "Bacterial Anatomy." *Symp. Soc. Gen. Microbiol.* **6:**261.

Marinozzi, V. 1963. "The Role of Fixation in Electron Microscopy Staining." *J. Royal Micr. Soc.* **81:**141–154.

———, and A. Gautier. 1962. "Fixations et Colorations. Etude des Affinités des Composants Nucléoprotéinques pour L'hydroxyde de Plomb et L'acetate d'Uranyle." *J. Ultrast. Res.* **7:**436–451.

McLean, R. C., and W. R. I Cook. 1952. *Plant Science Formulae* (Macmillan and Co. Ltd., London, England), 205 pages.

Mekler, L. B., S. M. Kilmenko, G. E. Dobrezov, V. K. Naumova, Y. P. Hoffman, and V. M. Zhdanov, 1964. "Cytochemical and Immunochemical Analysis at the Electron Microscope Level. Obtaining Contrasting Antibodies by Use of Iodine." *Nature* **203:**717–719.

REFERENCES

Mercer, E. H., and M. S. C. Birbeck. 1966. *Electron Microscopy. A Handbook for Biologists.* 2nd ed. (Blackwell Scientific Publications, Oxford, England), 102 pages.

Millonig, G. 1961a. "A Modified Procedure for Lead Staining of Thin Sections." *J. Biophys. Biochem. Cytol.* **11**:736–739.

———. 1961b. "Advantages of a Phosphate Buffer for OsO_4 Solutions in Fixation." *J. Appl. Phys.* **32**:1637.

Mollenhauer, H. H. 1959. "Permanganate Fixation of Plant Cells." *J. Biophys. Biochem. Cytol.* **6**:431–435.

———. 1963. "Plastic Embedding Mixtures for Use in Electron Microscopy." *J. Stain Tech.* **39**:111–114.

Moor, H. 1964. "Die Gefrier-Fixation lebender Zellen und ihre Anwendung in der Elektronmikroskopie." *Z. Zellforsch.* **62**:546–580.

———. 1969. "Freeze-Etching." *Inter. Rev. Cytol.* **25**:391–412.

Moses, M. J. 1964. "Application of Autoradiography to Electron Microscopy." *J. Histochem. Cytochem.* **12**:115–130.

Newman, S. B., E. Borysko, and M. Swerdlow. 1949. "Ultramicrotomy by a New Method." *J. Res. Natl. Bur. Stand.* **43**:183–199.

Palade, G. E. 1952. "A Study of Fixation for Electron Microscopy." *J. Exp. Med.* **95**:285–298.

Parsons, D. F. 1961. "A Simple Method for Obtaining Increased Contrast in Araldite Sections by Using Postfixation Staining of Tissues with Potassium Permanganate." *J. Biophys. Biochem. Cytol.* **11**:492–497.

Peachy, L. D. 1958. "Thin Sections I. A Study of Section Thickness and Physical Distortion Produced during Microtomy." *J. Biophys. Biochem. Cytol.* **4**:233–242.

Pearse, A. G. E. 1968. *Histochemistry. Theoretical and Applied.* Vol. 1. (Little, Brown and Company, Boston, Mass.), 759 pages.

Pease, D. C. 1964. *Histological Techniques for Electron Microscopy.* 2nd ed. (Academic Press, N.Y.), 381 pages.

——— and R. R. Baker. 1948. "Sectioning Techniques for Electron Microscopy Using a Conventional Microtome." *Proc. Soc. Exptl. Biol. Med.* **67**:470–474.

Pepe, F. A. 1961. "The Use of Specific Antibody in Electron Microscopy. I. Preparation of Mercury Labeled Antibody." *Jour. Biophys. Biochem. Cytol.* **11**:515–520.

Ramus, J. 1969. "Pit Connection Formation in the Red Alga *Pseudogloiophloea*." *J. Phycol.* **5**:57–64.

Rdzok, E. J. 1965. "The Outside Chamfering Mill—A Hand Tool for Rapid Preliminary Trimming of Blocks for Ultratomy." *Stain Tech.* **40**:371–372.

Reynolds, E. S. 1963. "The Use of Lead Citrate at High pH as an Electron Opaque Stain in Electron Microscopy." *J. Cell Biol.* **17**:208.

Ris, H. 1969. "Use of the High Voltage Electron Microscope for the Study of Thick Biological Specimens." *J. de Micr.* **8**:761–766.

Ryter, A., and E. Kellenberger. 1958. "L'Inclusion au Polyester pour L'Ultramicrotomie. *J. Ultrast. Res.* **2**:200–214.

REFERENCES

Sabatini, D. D., K. G. Bensch, and R. J. Barnett. 1962. "New Fixatives for Cytological and Cytochemical Studies," in *Fifth International Congress for Electron Microscopy (Phila.) Vol. II* (Academic Press, N.Y.), pages 1–3.

———. 1963. "Cytochemistry and Electron Microscopy. The Preservation of Cellular Ultrastructure and Enzymatic Activity by Aldehyde Fixation." *J. Cell Biol.* **17**:19–58.

Salpeter, M. M. and L. Bachmann. 1964. "Autoradiography with the Electron Microscope. A Procedure for Improving Resolution, Sensitivity, and Contrast." *J. Cell Biol.* **22**:469–477.

Sell, J. C., and W. W. McMaster, Jr. 1963. "Planning the Electron Microscope Suite." *J. Amer. Inst. of Architects.* May issue.

Singer, S. J. 1959. "Preparation of an Electron-Dense Antibody Conjugate." *Nature* **183**:1523–1524.

———, and A. F. Schick. 1961. "The Properties of Specific Stains for Electron Microscopy Prepared by the Conjugation of Antibody Molecules with Ferritin." *J. Biophys. Biochem. Cytol.* **9**:519–537.

Sjöstrand, F. S. 1967. *Electron Microscopy of Cells and Tissues. Vol. I. Instrumentation and Techniques* (Academic Press, N.Y.), 462 pages.

———, and R. F. Baker. 1958. "Fixation by Freeze-Drying for Electron Microscopy of Tissue Cells." *J. Ultrast. Res.* **1**:239–246.

Sorvall Inc., Ivan. 1965. "Free Break Method of Obtaining Glass Knives" (Ivan Sorvall Inc., Norwalk, Conn.), 7 pages.

———. 1967. "Thin Sectioning and Associated Techniques for Electron Microscopy," 2nd ed. (Ivan Sorvall Inc., Norwalk, Conn.), 123 pages.

Spurr, A. R. 1969. "A Low-Viscosity Epoxy Resin Embedding Medium for Electron Microscopy." *J. Ultrast. Res.* **26**:31–43.

Stachelin, L. A. 1968. "The Interpretation of Freeze-Etched Artificial and Biological Membranes." *J. Ultrast. Res.* **22**:326–347.

Sternberger, L. A., E. J. Donati and C. E. Wilson. 1963. "Electron Microscopic Study on Specific Protection of Isolated *Bordetella bronchiseptica* Antibody during Exhaustive Labeling with Uranium." *J. Histochem. Cytochem.* **11**:48–58.

Strangeways, T. S. P., and R. G. Canti. 1927. "The living Cell *In Vitro* as Shown by Dark-Ground Illumination and the Changes Induced in Such Cells by Fixing Reagents." *Quart. J. Micr. Sci.* **7**:1–4.

Thomas, R. J., and B. C. Maher. 1969. "A Simplified Protocol for Electron Microscopy. Processing of 'Pellet-able' Material." *Mikroskopie* **24**:296–298.

Thornton, P. R. 1968. *Scanning Electron Microscopy. Applications to Materials and Device Science* (Chapman and Hall, Ltd., London, England), 368 pages.

Von Ardenne, M. 1939. "Die Keilschnittmethode, ein Weg zur Herstellung von Mikrotomschnitten mit Weniger als 10^{-3} mm Stärke für Elektronmikroskopische Zwecke." *Zeitschr. f. wissensch. Mikroskopie und f. mik. Technik.* **58**:8–23.

Wardrop, A. B. 1964. "The Structure and Formation of the Cell Wall in Xylem," in M. H. Zimmerman (Ed.), *Formation of Wood in Forest Trees* (Academic Press, N.Y.), pages 84–134.

REFERENCES

Watson, M. L. 1958. "Staining of Tissue Sections for Electron Microscopy with Heavy Metals. II. Application of Solutions Containing Lead and Barium." *J. Biophys. Biochem. Cytol.* **4**:475–479.

Wischnitzer, S. 1962. *Introduction to Electron Microscopy.* (Pergamon Press, Inc., N.Y.), 132 pages.

———. 1970. *Introduction to Electron Microscopy.* 2nd ed. (Pergamon Press, Inc., N.Y.), 300 pages.

Wohlforth-Bottermann, K. E. 1956. "Die Entstehung, die Vermehrung und die Abscheidung geformter Sekrete der Mitochondrien von Paramecium" in F. S. Sjöstrand and J. Rhodin (Eds.), *Electron Microscopy Proc. Stockholm Conf. 1956* (Almquist and Wiksell, Uppsala, Sweden), pages 137–139.

Appendix I COMMERCIAL SOURCES AND CHEMICAL CHECKLIST

Beem Capsules I.1

Beem Inc., P. O. Box 132, Jerome Ave. Station, Bronx, New York 10468.

Electron Microscope Companies I.2

A. E. I.: Picker Nuclear, 1275 Mamaroneck Ave., White Plains, New York 10605.
Hitachi: Perkin-Elmer Corp., 8555 16th Street, Silver Springs, Maryland 20910.
JEM: Japan Electron Optics Laboratory Co., Ltd. (JEOLCO). 477 Riverside Avenue, Medford, Massachusetts 02155.
MAAK-1: Akashi Ltd., Tokyo, Japan.
MAC: Materials Analysis Co., 1060 East Meadow Circle, Palo Alto, California 94303.
Philips (Norelco): Philips Electronic Instruments, 750 South Fulton Ave., Mount Vernon, New York 10550.
RCA: Radio Corporation of America, Scientific Instruments, Camden, New Jersey 08102. (As of June, 1969, the production of the RCA electron microscope was sold to Forgflo Corp., Waltham Industries Subsidiary, Sunbury, Pennsylvania 17801.)
Siemens: Siemens of America, Inc., 350 Fifth Avenue, New York City, New York 10001.
UEMB-100K: V/O Mashpriborintorg, 32/34 Smolenskaja Sq., Moscow, G200, USSR.
Ziess EM 9S: Carl Ziess Inc., 444 Fifth Ave., New York City, New York 10018.

General Electron Microscope Supply Houses I.3

Beem Inc., P.O. Box 132, Jerome Ave. Station, Bronx, New York 10468.
Electron Microscopy Sciences (EMS), Box 251, Fort Washington, Pennsylvania 19034.
C. W. French, Inc., 58 Bittersweet Lane, Weston, Massachusetts 02193.
E. F. Fullam, Inc., P.O. Box 444, Schenectady, New York 12301.
Ladd Research Industries, Inc., P.O. Box 901, Burlington, Vermont 05401.
Polysciences, Inc., Paul Valley Industrial Park, Warrington, Pennsylvania 18976.

I SOURCES AND CHEMICAL CHECKLIST

Vaughn Electron Microscopy Supplies, Inc., 2176 Dunn Road, Memphis, Tennessee 38114.
Western Instrument Scientific Equipment (WISE), Route 2, Box 1365, Vacaville, California 95688.

I.4 Laboratory Chemicals—A Checklist

acetic acid (CH_3COOH)
acetone ($CH_3 \cdot CO \cdot CH_3$)
acrylic aldehyde, or acrolein ($CH_2 : CH \cdot CHO$)
agar
albumen
amyl acetate ($CH_3COOC_5H_5$)
azure II
barium permanganate ($Ba(MnO_4)_2$)
basic fuchsin
benzene (C_6H_6)
tert-butyl alcohol (($CH_3)_3COH$)
calcium chloride ($CaCl_2$)
celluloid plastic sheets, or cellulose acetate
charcoal, bone, activated
chloroform ($CHCl_3$)
crystal violet
ethanol, 100% ($CH_3 \cdot CH_2OH$)
ethanol, 95% ($CH_3 \cdot CH_2OH$)
ethylene dichloride, or dichlorethane (CH_3CHCl_2)
ethylene glycol ($CHOH_2CH_2OH$)
formvar
glucose ($C_6H_{12}O_6$)
glutaraldehyde, or pentonedial ($CHO(CH_2)_3CHO$)
gum mastic
glycerol ($CHOH(CH_2OH)_2$)
hydrochloric acid (HCl)
lead acetate ($Pb(C_2H_3O_2)_2$)
lead acetate, basic ($Pb(C_2H_3O_2)_2Pb(OH)_2H_2O$)
lead hydroxide ($Pb(OH)_2$)
lead nitrate ($Pb(NO_3)_2$)
lead oxide, mono (PbO)
methanol ($CH_3 \cdot OH$)
methylene blue
molybdic acid ($H_2MoO_4 \cdot H_2O$)
osmic acid, or osmium tetroxide (OsO_4)

paraffin wax
paraformaldehyde (($CH_2O)_x \cdot _xH_2O$)
parlodion
phosphotungstic acid ($P_2O_5 \cdot 24WO_3 \cdot nH_2O$)
polyvinyl alcohol
potassium chloride (KCl)
potassium dichromate ($K_2Cr_2O_7$)
potassium hydroxide (KOH)
potassium permanganate ($KMnO_4$)
potassium phosphate, monobasic (K_2HPO_4)
potassium phosphate, dibasic (KH_2PO_4)
potassium sodium tartrate ($KNaC_4H_4O_6 \cdot 4H_2O$)
propylene oxide, or trimethylene oxide ($CH_2 \cdot CH_2 \cdot CH_2O$)
ruthenium red ($Ru_2(OH)_2Cl_4 \cdot 7NH_3 \cdot 3H_2O$)
sodium (metal) (Na)
sodium acetate ($NaC_2H_3O_2$)
sodium barbital, or sodium veronal ($NaC_8H_{11}N_2O_3$)
sodium bicarbonate ($NaHCO_3$)
sodium borate (tetra), ($Na_2B_4O_7 \cdot 10H_2O$)
sodium cacodylate ($Na((CH_3)_2AsO_2) \cdot 3H_2O$)
sodium chloride (NaCl)
sodium citrate ($NaC_6H_5O_7 \cdot 2H_2O$)
sodium hydroxide (NaOH)
sodium phosphate, monobasic ($NaH_2PO_4 \cdot H_2O$)
starch ($C_6H_{10}O_5)_x$
sucrose ($C_{12}H_{22}O_{11}$)
toluene, or methyl benzene ($C_6H_5 \cdot CH_3$)
toluidine blue ($CH_3 \cdot C_6H_4 \cdot NH_2$)

2,4,6 trimethyl pyridine, or s-collidine $((CH_3)_3 \cdot C_5H_2N)$
uranyl acetate $(UO_2(C_2H_2O_2)_2 \cdot 2H_2O)$
uranyl nitrate $(UO_2(NO_3)_2 \cdot 6H_2O)$
vanadyl sulfate $(VOSO_4)$
xylene $(C_6H_4(CH_3)_2)$

Plastics I.5

Accelerator B (N-Benzylmethylamine): R. P. Cargille Laboratories, Inc., 33 Village Park Road, Cedar Grove, New Jersey 07009.

Accelerator S-1 (DMAE; dimethylaminoethanol): Pennsalt Chemical Corp., Three Penn Center, Philadelphia, Pennsylvania 19102. Also available as part of a plastic kit from Polysciences, Inc., Paul Valley Industrial Park, Warrington, Pennsylvania 18976.

Araldite 506 (714): CIBA Products, Fair Lawn, New Jersey 07410.

Araldite 6005 (same as Epoxy resin Araldite 502): R. P. Cargille Laboratories, Inc., 33 Village Park Road, Cedar Grove, New Jersey 07009.

BDMA (Benzyldimethylamine): Polysciences, Inc., Paul Valley Industrial Park, Warrington, Pennsylvania 18976.

Benzoyl Peroxide: Polysciences, Inc., Paul Valley Industrial Park, Warrington, Pennsylvania 18976.

Cardolite NC513: Minnesota Mining and Manufacturing Co., Irving Chemical Division, Chemical Products Group, 500 Doremus Avenue, Newark, New Jersey 07105.

Cobalt naphthenate: Polysciences, Inc., Paul Valley Industrial Park, Warrington, Pennsylvania 18976.

DDSA (Dodecenyl succinic anhydride): R. P. Cargille Laboratories, Inc., 33 Village Park Road, Cedar Grove, New Jersey 07009.

DER 7361 (Diglycidyl ether of propylene glycol): Dow Chemical Corporation, Plastic Sales, Midland, Michigan 48640. Also available as part of a kit from Polysciences, Inc., Paul Valley Industrial Park, Warrington, Pennsylvania 18976.

Dibutyl-phthalate: R. P. Cargille Laboratories, Inc., 33 Village Park Road, Cedar Grove, New Jersey 07009.

DMP-30 (2,4,6 dimethylamino-methylphenol): Rohm and Haas, Washington Square, Philadelphia, Pennsylvania 19106.

Epon 812: Shell Chemical Corp., Plastics Division, New York City, New York 10006.

Maraglas 655: Polysciences, Inc., Paul Valley Industrial Park, Warrington, Pennsylvania 18976.

Methyl methacrylate: Rohm and Haas, Washington Square, Philadelphia, Pennsylvania 19106.

N-butyl methacrylate: Rohm and Haas, Washington Square, Philadelphia, Pennsylvania 19106.

NMA (Nadic methyl anhydride): National Aniline Division of Allied Chemical Corp., 40 Rector Street, New York City, New York 10006. Also available as part of a kit from Polysciences, Inc., Paul Valley Industrial Park, Warrington, Pennsylvania 18776.

I SOURCES AND CHEMICAL CHECKLIST

NSA (Nonenyl succinic anhydride): Humphrey Chemical Co., Devine Street, North Haven, Connecticut 06473 (specify "refined for electron microscopy"). Also available as part of a kit from Polysciences, Inc., Paul Valley Industrial Park, Warrington, Pennsylvania 18976.

Tert-butyl perbenzoate: Polysciences, Inc., Paul Valley Industrial Park, Warrington, Pennsylvania 18976.

UNOX Epoxide 206 (ERL 4206): Union Carbide Corp., 270 Park Avenue, New York City, New York 10017.

Vestopal W: Morton M. Jager, Vésenaz, Geneva, Switzerland. Also available as part of a kit from Polysciences, Inc., Paul Valley Industrial Park, Warrington, Pennsylvania 18976.

I.6 Diamond Knives

Ernst Leitz, Wetzlar, Germany.

Instituto Venezoleno de Investigaciones Cientificos (I.V.I.C), Apartitado 187, Caracas, Venezuela.

Instrument Products Division, E. I. duPont de Nemours & Co., Inc., Wilmington, Delaware 19898.

Ladd Research Industries, Inc., P.O. Box 901, Burlington, Vermont 05401.

Rondikn Corporation, 2003 Kalia Road, PH 7, Honolulu, Hawaii 96815.

I.7 Knife Makers

LKB Knife Maker 7800B: LKB Instruments, Inc., 12221 Parklawn Drive, Rockville, Maryland 20852.

Sunkay Messer: C. W. French, Inc., 58 Bittersweet Lane, Weston, Massachusetts 02193.

I.8 Ultramicrotomes

Cambridge Ultramicrotome: Cambridge Instrument Co., Ltd., London S.W. L, England.

Leitz Ultramicrotome: E. Leitz Inc., 468 Park Ave. South, New York City, New York 10016.

LKB Ultratome and Ultratome III: LKB Instruments, Inc., 12221 Parklawn Drive, Rockville, Maryland 20852.

Reichert OmU2: Heinalsen Hauptstrasse 219 A-1171, Wien, Austria.

SI-RO-FLEX Ultramicrotome: Schyco Scientific Division of Schueler & Co., 75 Cliff Street, New York City, New York 10038.

Sorvall Porter Blum MT-1, MT-2: Ivan Sorvall, Inc., Norwalk, Connecticut 06862.

Plastic Film Materials   I.9

Formvar: Shawinigan Products Co., 350 Fifth Ave., New York City, New York 10001.
Parlodion (Collodion): Mallinckrodt Chemical Works, St. Louis, Missouri 63100.

Specimen Grid Supply Houses   I.10

(See also Sections I.2 and I.3.)

Grids for Microscopy (GFM), 107 N. Brookside Drive, Dallas, Texas 75214.
Mason and Morton, Ltd., Fir Tree House, Headstone Drive, Middx., England.
Ted Pella Co., P.O. Box 606, Altadena, California 91001.
Whittaker Corp., Gencom Division, 80 Express Street, Plainview, New York 11803.

Special Sources of Osmium Tetroxide   I.11

(See also Section I.3.)

United Mineral and Chemical Corp., 129 Hudson St., New York City, New York 10013.
Stevens Metallurgical Corp., 342 Madison Ave., New York City, New York 10017.

High Vacuum Evaporator Companies   I.12

Bendix-Balzers Vacuum Inc., 1645 St. Paul Street, Rochester, New York 14621.
Denton Vacuum Inc., 8 Fellowship Road, Cherry Hill, New Jersey 08033.
Edwards High Vacuum Inc., 3279 Grand Island Blvd., Grand Island, New York 14072.
Kinney Vacuum Division, The New York Brake Company, 3529 Washington Street, Boston, Massachusetts 02130.
Ladd Research Industries Inc., P.O. Box 901, Burlington, Vermont 05401.
Mikros Inc., 7634 S.W. Capital Highway, Portland, Oregon 97200.

Appendix II CHEMISTRY OF EPOXY RESINS

Introduction   II.1

Epoxy resins are the most commonly used group of embedments in electron microscopy today. Because of the rather poorly understood process of polymerization and the number of components in formation of epoxy plastics, this brief account of the chemistry of epoxy resins is included here rather than in the chapter on plastics.

As stated in chapter 5, most epoxy plastics are made from mixtures of four components:

1. epoxy resin
2. hardener
3. plasticizer
4. accelerator

Each component will be discussed separately.

Epoxy Resin   II.2

The epoxy resin itself, whether epon or araldite, is characterized by the presence of oxirane groups and is usually the result of condensation polymerization of epichlorohydrin and bisphenal containing monomers in the presence of alkali. Cargille Laboratories (1962) gives the following chemical synthesis of an epoxy resin (araldite) from an epichlorohydrin and bisphenal A* (2:2 bis (4-hydroxyphenyl) propane):

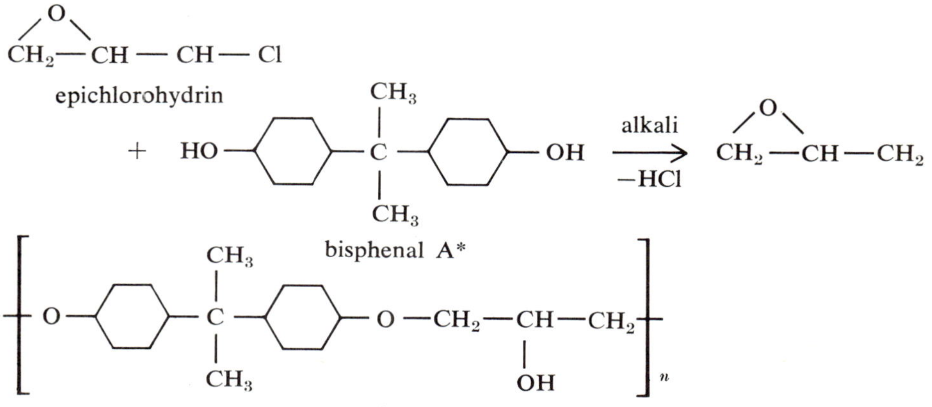

$$\underset{\text{epichlorohydrin}}{CH_2\text{—}CH\text{—}CH\text{—}Cl} + HO\text{—}\phi\text{—}\underset{CH_3}{\overset{CH_3}{C}}\text{—}\phi\text{—}OH \xrightarrow[-HCl]{\text{alkali}} CH_2\text{—}CH\text{—}CH_2$$

bisphenal A*

$$\left[\text{—}O\text{—}\phi\text{—}\underset{CH_3}{\overset{CH_3}{C}}\text{—}\phi\text{—}O\text{—}CH_2\text{—}\underset{OH}{CH}\text{—}CH_2\text{—} \right]_n$$

epoxy resin

(*continued*)

II CHEMISTRY OF EPOXY RESINS

$$O-\underset{\underset{CH_3}{|}}{\overset{\overset{CH_3}{|}}{C}}-\bigcirc-O-CH_2-CH-CH_2 \quad (\text{epoxide})$$

where n may vary from 0 to perhaps 13 (mol. wt. : 340 to 4000).

These epoxy resins range from relatively low viscosity liquid materials to high melting point solids, depending on their molecular weights. The lower viscosity, lower molecular weight epoxy materials of this type are selected as embedding media for tissues since their higher diffusion rates promise more effective tissue penetration.

Another epoxy resin is the very low molecular weight ERL-4206 (vinyl cyclohexene dioxide), which is a cycloaliphatic diepoxide (Spurr (1969)). The formula is

(structural diagram of vinyl cyclohexene dioxide)

Apparently this compact diepoxide structure yields linear polymers that are highly cross linked, and consequently are very thermostable. The major attribute, however, appears to be the very low molecular weight (140.18) making it a clear, mobile liquid with a specific gravity of 1.10 at 20° C. The commercial title is UNOX Epoxide 206 and is obtained from Dow Chemical Corporation.

Other commercially available epoxy resins include those prepared by the condensation polymerization of epichlorohydrin with various aliphatic and aromatic dioles and polyoles (bi- and poly-functional alcohols). Epoxy materials are also prepared commercially by the peracetic acid oxidation of aliphatic and aromatic dienes (molecules containing two sets of double bonds).

II.3 Hardeners

Hardeners are mixed with the epoxy resins to cause cross linking. Epoxy resins are compounds containing more than one terminal epoxy group and they can be cured or hardened to form insoluble thermostable solids by poly-addition with cross-linking agents containing active hydrogen atoms, or by self-polymerization in the presence of catalysts. Such curing agents or hardeners may be divided into the following four categories:

1. Acid hardeners—polycarboxylic acids, dicarboxylic acid anhydrides, polyesters with free carboxyl groups, sulfonic and phosphoric acids, and polysulphides.
2. Amine hardeners—aliphatic polyamines and amino-polyamides.
3. Aldehyde-condensation products—phenol, and melamine- and urea-formaldehyde resins.
4. Inorganic or organic-metallic compounds—boron trifluoride, titanic acid esters, and aluminum alcoholate.

The cure of epoxy systems by acid anhydrides has received the most attention for tissue embedment media. Dodecanyl succinic anhydride (DDSA) is a light yellow, clear, viscous oil having a calculated molecular weight of 266. Its molecular formula (Cargille (1962)) is

$$C_{12}H_{23} - HC - C(=O) - O - C(=O) - CH_2$$

The hardener recommended by Spurr (1969) is very similar to that described by Cargille above. It is NSA (nonenyl succinic anhydride), another branched alkenyl succinic anhydride based on tripropylene with a molecular weight of 224. The general formula is

$$CH_3-CH(CH_3)-CH_2-CH(CH_3)-CH_2-CH_2-CH_2-CH-C(=O)-O-C(=O)-CH_2$$

This hardener has a lower viscosity than DDSA (117 cP vs. 290 cP) and is water-white in color.

Plasticizers II.4

Plasticizers are used to improve the cutting qualities of the epoxy resin-hardener mixtures. These materials are basically nonreactive in polymerization, act as a filler, and impart higher flexibility and softness to the cured mixture.

Dibutyl-phthalate is one such plasticizer that also reduces the viscosity of the cured mixture. The formula is

II CHEMISTRY OF EPOXY RESINS

$$\text{benzene ring with } -CO_2H, -CO_2H$$

Spurr (1969) recommends the use of an epoxy resin DER 736 (diglycidyl ether of propylene glycol) as a plasticizer because of its low viscosity. The formula is

$$CH_2\overset{O}{-}CH-CH_2-O-\left[\underset{\underset{n=4}{\big|}}{\overset{CH_3}{\big|}}CH-CH_2-O-CH_2\right]-CH_2-CH\overset{O}{-}CH_2$$

The molecular weight is 380 (average) with a specific gravity of 1.14 at 25° C.

II.5 Accelerators

Accelerators are used to promote curing of the plastic mixture, and apparently are not involved themselves in polymerization. Thus, they are considered to be catalysts. One such accelerator is n-benzyl dimethylamine,

$$\text{cyclohexyl}-CH_2-N\underset{CH_3}{\overset{CH_3}{\diagup\!\!\diagdown}}$$

It should be recognized that the quantity of amine accelerator is critical. Small deviations from the recommended amounts can produce great changes in pot life and in the properties of the final product cured under given conditions.

Spurr recommends another accelerator, X-1 (dimethylaminoethanol) or DMAE, which is one of the alkyl alkanol amines and its structure is

$$\underset{CH_3}{\overset{CH_3}{\diagdown\!\!\diagup}}N-CH_2-CH_2-OH$$

It has the typical tertiary amine position and curing effects for epoxy resins with a specific gravity of 0.89 and a molecular weight of 89 to 93. It permits rapid curing at 70° C.

Warning: Breathing the vapor of an amine containing accelerator is harmful and the liquid may cause eye injury and skin burns. Some epoxy materials and amine hardeners may cause dermatitis in sensitive persons. In all cases good ventilation, cleanliness, and the use of gloves or barrier cream is recommended.

Curing Mechanisms of Anhydride-Epoxy Systems II.6

The following sequence of reactions has been proposed by Fisch and Hofmann (1956) to elucidate the heat cure of epoxy resins by anhydride hardeners.

1. The first step is the formation of carboxyl-bearing monoester by a reaction between the anhydride of the hardener and secondary alcohol groups present on the epoxy resins. Generalizing from Fisch and Hofmann's work with phthalic anhydride the following reaction should take place:

$$\underset{\text{secondary alcohol}}{HC-OH} + \underset{\text{anhydride}}{O=C\overset{O}{\diagup\diagdown}C=O} \longrightarrow \underset{\text{carboxyl-bearing monoester}}{HC-O-\underset{O}{\overset{O}{\|}}C\quad \underset{}{\overset{O}{\|}}C-OH}$$

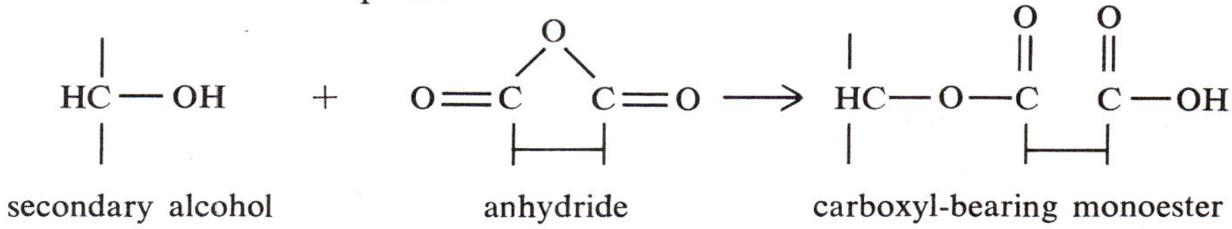

2. The next step in the mechanism proposed by Fisch and Hofmann is the reaction of the carboxyl-bearing monoester with the terminal epoxide group of another epoxy resin molecule. Again generalizing from the work with phthalic anhydride the following reaction should take place:

$$\underset{\text{carboxyl-bearing monoester}}{HC-O-\overset{O}{\overset{\|}{C}}-\overset{O}{\overset{\|}{C}}-OH} + \underset{\text{epoxy}}{CH_2-CH-\overset{O}{\diagup\diagdown}}$$

$$\longrightarrow HC-O-\overset{O}{\overset{\|}{C}}\quad \overset{O}{\overset{\|}{C}}-O-CH_2-\underset{\text{diester bridge}}{CH-}\overset{OH}{\overset{|}{}}$$

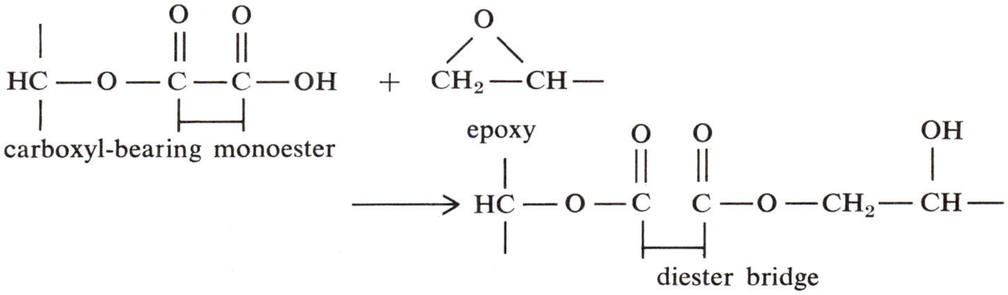

Thus two epoxy molecules are joined by a diester bridge. Since each resin molecule possesses another terminal epoxy group on its free end, and since a new secondary alcohol is regenerated by this reaction, the macromolecule has ample opportunity for further growth, or polymerization with a hardener.

3. In some cases where the epoxy materials are without hydroxyl groups, it is thought that small amounts of water or carboxylic acid are present, which activate the epoxide group thus:

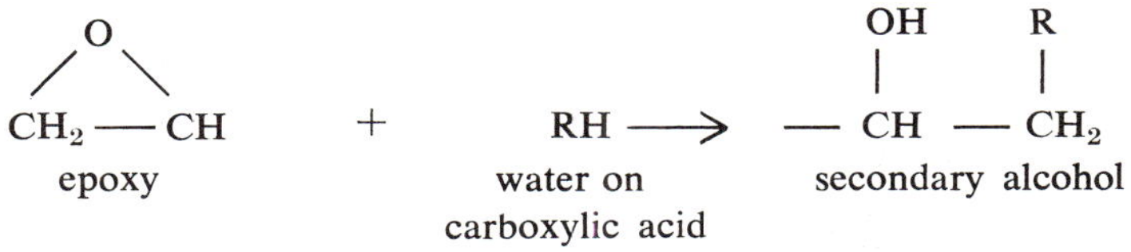

$$\underset{\text{epoxy}}{CH_2\overset{O}{\diagup\diagdown}CH} + \underset{\substack{\text{water on}\\ \text{carboxylic acid}}}{RH} \longrightarrow \underset{\text{secondary alcohol}}{-CH\overset{OH}{\overset{|}{}}-CH_2\overset{R}{\overset{|}{}}}$$

II CHEMISTRY OF EPOXY RESINS

With a trace of secondary alcohol present, formation of carboxyl-bearing monoester can take place as described in steps 1 and 2. Since the carboxyl-bearing monoester goes on to react with epoxide (see above) with the regeneration of secondary alcohol groups, these reactions can be self-sustaining. However, the polyglycidyl ether used in epoxide resins should have at least a moderate degree of precondensation.

4. Fisch and Hofmann (1956) have also proposed the occurrence of a fourth reaction, the self-polymerization of epoxy resins under the catalytic influence of anhydride or carboxylic acid. This occurs by the addition of a hydroxyl group of one molecule to the epoxide group of a second molecule thus:

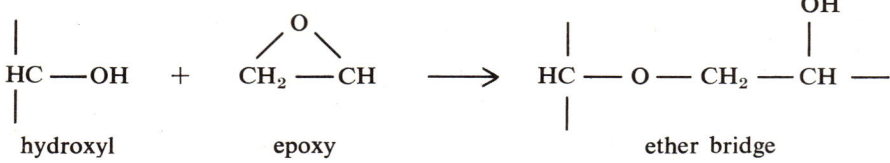

Thus the cross-linking of epoxy resins with anhydrides is based not only on diester bridges, but on ether bridges as well. In summary, the four steps of polymerization of epoxy resins and hardeners described by Fisch and Hofmann (1956) are presented showing both the occurrence of cross linkage and self-polymerization.

It has been noted that amines are potent accelerators for epoxy-anhydride systems. Since only small amounts are required and since the essential nature of the epoxy-anhydride reaction appears to be unchanged, it is felt that the amine functions as an accelerator rather than a reagent.

Appendix III   SAMPLE PREPARATION SCHEDULE

Introduction   III.1

The outline of procedures given below has been found to yield good results regardless of the material in question. Bacterial fixation and handling of any unicellular material requires special techniques. For this, the investigator is referred to the osmium fixation of Kellenberger *et al.* (1958, chapter 2) and to the techniques described for pre-embedding of particulate specimens (chapter 4). Furthermore, the actual handling of higher animal tissue may require perfusion or *in situ* preliminary fixation. Consequently, the reader will have to make many decisions on preliminary handling.

Fixation   III.2

In order to obtain an accurate view of the cell structure it is recommended that at least two and possibly three chemical fixatives are used. The recommended fixatives and buffers for beginning students should be:

1. osmium tetroxide, 1%, in Millonig's phosphate buffer (chapter 2);
2. glutaraldehyde buffered in cacodylate followed by osmium tetroxide postfixation in either the same buffer or in Millonig's phosphate buffer (chapter 2); and
3. $KMnO_4$ in Luft's veronal-acetate buffer (chapter 2).

The factors affecting fixation such as length of fixation, pH, temperature, and tonicity are discussed in the introduction to chapter 2 and suggested times for fixation are given for each. After the investigator has examined the results of fixation using a recommended duration, temperature, and buffer, it will be possible to determine whether or not changes should be made in the procedures.

Some general points are worthy of recitation. The investigator should first read the section on preparation of tissue for fixation and have all ingredients ready before attempting to actually carry out the procedure. Any inordinate delay between killing or removing tissue from the organism and actual fixation will probably result in some cell disruption. Rinsing between primary and postfixation is very important. A good rule of thumb is not less than 1 hr with three to four changes of fresh buffer (some investigators recommend overnight rinsing). Throughout rinsing handle the tissue carefully, do not use tweezers but rather the flattened ends of toothpicks (pick up the tissue only when necessary to avoid mechanical damage).

III SAMPLE PREPARATION SCHEDULE

III.3 Dehydration

The recommended schedule is dehydration in ethanol followed by propylene oxide (chapter 3, schedule 1). As explained in the recipe for the schedule, there are two suggested sequences, one is using a complete series of 10% steps in ethanol with about 15 min at each step, while the other sequence is much more rapid, moving from 50% to 70% to 100% ethanol and then directly into propylene oxide, all done in the cold. The investigator should probably try both series to determine whether the rapid procedure will be helpful. Such symptoms as cytoplasmic shrinkage or wall-vacuole collapse should be looked for and if found a slower progression with smaller dehydration increases is recommended. Again, do not handle the tissue, simply pour off the old solution, and add the new. Allow a slight amount of solution to remain at the bottom of the vial to cover the specimens to prevent dessication during solution changing.

III.4 Embedding

There are a number of good plastics available. For simplicity only two are recommended here: Spurr's low viscosity epon (chapter 5) and Mollenhauer's epon-araldite mixture 2 (chapter 5). With each, the author has received consistently good results in embedding. The steps for embedding are given for each of the mixtures in chapter 5. Again, the investigator should have all solutions and mixtures made up before proceeding from 100% ethanol (for Spurr's epon) or propylene oxide (for Mollenhauer's epon-araldite). Tissue handling should be done with toothpicks to prevent mechanical damage. Embedding probably should be done first in capsules using capsule holders. Flat embedding may follow in a second embedding or parallel with capsule embedding.

III.5 Block Trimming and Grid Preparation

As recommended in chapter 6, the final block size should not be more than 0.2 mm on the longest side of the trapezoid. It is wise to pick up sections on both naked (uncoated) 300-mesh grids and formvar coated 200- or 100-mesh grids. Formvar is recommended as a coating plastic because of its greater stability under the electron beam and because it does not require an additional coating of carbon. The recommended procedure for coating grids is Method 1 as described in chapter 8 using formvar.

III.6 Sectioning and Staining

As explained in chapter 7, the ultramicrotome area should be set with all needed material for sectioning. Once the sections have been picked up on a grid, staining should follow as soon as possible. A standard procedure recommended is the double stain method

SECTIONING AND STAINING III.6

using uranyl acetate and lead citrate. The length of time should be around 30 min and 15 min, respectively. The reader is cautioned not to attempt staining more than three grids in a single container when working with any lead stain. This will prevent too much exposure to the air each time a grid is removed and in turn will cut down on lead carbonate formation.

Appendix VI
LOCALIZATION TECHNIQUES IN ELECTRON MICROSCOPY

Introduction IV.1

A number of specialized techniques in electron microscopy have been developed in the past ten years to permit localization of enzymatic, synthetic, or protein sites at the subcellular level. Detailed descriptions of these highly technical procedures are beyond the scope of this text, but a brief introduction to these procedures with references is indeed desirable.

Immunochemical "Staining" IV.2

Antigen-antibodies, commonly used in immunochemical studies, have been employed in electron microscopy (Glauert (1961); Pearse (1968)). Because the antibodies are not distinctive in ultrathin sections, they must be labeled with some electron dense material. Ferritin, a protein which consists of about 20% iron, is frequently used by coupling it covalently with an antibody. The tetrahedral latticework consisting of ferric hydroxide-phosphate micelles within the protein shell of ferritin is quite distinctive. Thus, the electron dense particles of ferritin can be easily seen and recognized in the electron microscope (Singer (1959); Singer and Schick (1961)). By bonding the ferritin molecule to various antibodies, and introducing the antibody into the tissue, the investigator can determine the site of the antigen in the cell.

In addition to iron, antibodies have also been labeled with other electron dense elements such as mercury (Pepe (1961)), uranium (Sternberger *et al.* (1963)), and iodine (Mekler *et al.* (1964)).

Enzyme Localization IV.3

Enzymes can be localized in histochemical studies by allowing electron dense particles to be deposited in the cell at the site of an enzymatic reaction. This procedure permits correlation of fine structure with function using prepared ultrathin sections in an electron microscope. These controlled chemical procedures are complex and only a brief description can be given here (see the extensive discussion of Pearse (1968)).

Preparation of the specimen for enzyme location at the electron microscope level must insure that enzymatic activity survives in the

IV   LOCALIZATION TECHNIQUES

fixed, dehydrated, and embedded cell. Consequently, aldehydes are usually recommended as fixatives (Holt and Hicks (1961b); Sabatini, Bensch, and Barnett (1963)). After fixation, the tissue is usually frozen and cut into thin (30 μm) sections on a cryostat. These sections are then "stained" by incubating them in a suitable medium. The medium will depend on the enzyme to be localized. For example, lead sulphide can be deposited in tissue by hydrolysis of thioacetate acid in the presence of lead ions (Pearse (1968)). The lead sulphide particles, being electron dense, will then indicate the site of perhaps the enzyme esterase. Following incubation in the medium, the 30 μm thick, frozen sections are dehydrated, embedded, and thin-sectioned for electron microscopy. These sections of cells are then examined for sites of deposition of a marker such as lead sulphide.

IV.4   Autoradiography

Chemical constituents in the cell can be localized by photographically recording the position of radioactive material which has been introduced into the specimen. Autoradiography for electron microscopy has become a very detailed and well-used technique for localization of cell wall constituents in plants, nuclear constituents of cells, as well as reserve foods in cells (Caro et al. (1962); Moses (1964); Salpeter and Bachman (1964)).

The procedure, although quite technical in method, is simple to comprehend. The organism is exposed to a labeled compound which is incorporated into constituents of cells or cell walls. After fixation, dehydration, and embedding, ultrathin sections of the specimen containing radioactive material are coated with a thin layer of a liquid photographic emulsion (Koehler et al. (1963); Bourne and Cole (1969)). These coated sections are then stored in the dark for a period ranging from a few days to a number of months. The length of time for storage will depend on the sensitivity of the emulsion and the intensity of radiation. The radiation emitted during decay of the labeled material reduces silver bromide crystals which lie over the site of labeled material. Upon developing, the affected silver bromide crystals are further reduced, and silver grains are formed over the sites of radioactive decay.

Because of the high resolving power and depth of field of the electron microscope, the electron dense, silver grains are viewed as lying over sites of incorporation of the radioactive substance. Such techniques have been used for localization of lignin and cell wall precursors and nucleic acid precursors in nuclei formation.

Appendix V GENERAL REFERENCE LIST

The following list of general references is offered as a selection of supporting texts and major articles in the field of electron microscopy. The comments following some of the references are notes regarding possible uses.

Bahr, G. F., and E. H. Zeitler (Ed.). 1965. *Quantitive Electron Microscopy. Proceedings, Symposium on Quantitive Electron Microscopy, Washington, D. C., 1964* (The Williams & Wilkins Co., Baltimore, Md.).

Bakish, R. 1962. *Introduction to Electron Beam Technology* (John Wiley & Sons, Inc., N.Y.).

Beeching, R. 1946. *Electron Diffraction* (Methuen, London, England).

Birks, L. 1963. *Electron Probe Microanalysis* (Interscience Publications, John Wiley & Sons, Inc., N.Y.).

Breese, S. S., Jr. (Ed.). 1962. *Fifth International Congress for Electron Microscopy;* Vol. I: *Non-Biology;* Vol. II: *Biology* (Academic Press, N.Y.). [Detailed specific papers given at the Congress in Philadelphia.]

Burton, E. F., and W. H. Kohl. 1942. *The Electron Microscope* (Van Nostrand, Reinhold, N.Y.). [A simple, easy-to-read introduction to the electron microscope.]

Causey, G. 1962. *Electron Microscopy: A Textbook for Students of Medicine and Biology* (The Williams & Wilkins Co., Baltimore, Md.).

Clark, G. L. 1961. *Encyclopedia of Microscopy* (Van Nostrand, Reinhold, N.Y.).

Cosslett, V. E. 1951. *Practical Electron Microscopy* (Academic Press, N.Y.). [A book dealing with techniques; some are early techniques that are not currently used.]

Donaldson, P. E. K. 1959. *Electronic Apparatus for Biological Research* (Academic Press, N.Y.).

Fischer, R. B. 1954. *Applied Electron Microscopy* (Indiana University Press, Bloomington, Ind.). [A very useful reference for biologists on the workings of an electron microscope; some of the information is not currently applicable.]

Grivet, P., with collaboration of N. Y. Bernard *et al.;* revised by A. Septiu; translated by P. W. Hawkes. 1965. *Electron Optics* (Pergamon Press, Inc., N.Y.). [Part I covers electron lenses, part II, cathode ray tubes, microscopes, etc.; and chapters 15 through 20 are on electron microscopy.]

Hall, C. B. 1966. *Introduction to Electron Microscopy* (McGraw-Hill Book Company, N.Y.). [A technical discussion of the electron microscope.]

V GENERAL REFERENCE LIST

Haine, M. E. 1961. *The Electron Microscope* (Interscience Publications, John Wiley & Sons, Inc., N.Y.). [A mathematical treatise on electron microscope optics with a detailed discussion of the practical applications; very technical.]

Harris, R. (Ed.). 1962. *The Interpretation of Ultrastructure* (Academic Press, N.Y.). [Individual papers that deal with all phases of cell structure; a good reference for biologists.]

Hawley, G. G. 1945. *Seeing the Invisible: The Story of the Electron Microscope* (Alfred A. Knopf, Inc., N.Y.). [A simplified discussion of the electron microscope and its uses.]

Juniper, B. E., G. C. Cox, A. J. Gilchrist, and P. R. Williams. 1970. *Techniques for Plant Electron Microscopy* (Blackwell Scientific Publications, Oxford, England), 108 pages. A brief recipe book on techniques.

Kay, D. (Ed.). 1961. *Techniques for Electron Microscopy*. 2nd ed. (Blackwell Scientific Publications, Oxford, England). [A detailed, technical account of electron microscope methods. Excellent as a reference for the improvement of general techniques since each chapter written by experts in the field.]

Mercer, E. H., and M. S. C. Birbeck. 1966. *Electron Microscopy, A Handbook for Biologists*. 2nd ed. (Blackwell Scientific Publications, Oxford, England). [A recipe book of biological techniques in electron microscopy.]

Pease, D. C. 1964. *Histological Techniques for Electron Microscopy* (Academic Press, N.Y.). [Easy to read with a thorough handling of electron microscopy techniques in biology.]

Siegel, B. M. (Ed.). 1964. *Modern Developments in Electron Microscopy* (Academic Press, N.Y.). [A cross section of physical and biological techniques for electron microscopy.]

Sjöstrand, F. S. 1967. *Electron Microscopy of Cells and Tissues. Vol. I. Instrumentation and Techniques* (Academic Press, N.Y.). [A detailed account of the microscope and biological preparative techniques which are strongly oriented toward higher animals.]

Sorvall Inc., Ivan. 1965. *Thin Sectioning and Associated Techniques for Electron Microscopy*. 2nd ed. (Ivan Sorvall, Inc., Norwalk, Conn.). [A small paperback giving some good information on sectioning.]

Thornton, P. R. 1968. *Scanning Electron Microscopy. Applications to Materials and Device Science* (Chapman and Hall, Ltd., London, England). [The first book out on this exciting new field.]

Wischnitzer, S. 1970. *Introduction to Electron Microscopy*. 2nd ed. (Pergamon Press, Inc., N.Y.) [A simplified discussion of the theory and optics of the electron microscope and a rationale of the present-day biological techniques. The first section is essentially that of Wischnitzer's earlier book on electron microscopy (Wischnitzer, 1962, in references), while the latter portion presents descriptions of the procedures used in preparation of biological material but no actual techniques.]

Wycoff, R. W. G. 1949. *Electron Microscopy: Techniques and Applications* (Interscience Publications, John Wiley & Sons, Inc., N.Y.). [An out-of-print, useful text.]

Wycoff, R. W. G. 1958. *The Work of the Electron Microscope.*

GENERAL REFERENCE LIST V

(Yale University Press, New Haven, Conn.) [An elementary account of the electron microscope, its optics and accessories. A good introduction for laymen.]

Zworykin, V. K., *et al.* 1948. *Electron Optics and the Electron Microscope* (John Wiley & Sons, Inc., N.Y.). [A basic reference in electron microscopy.]